The Theory of Technology Evolution

Laurie Thomas Vass

The Great American Business & Economics Press
GABBY Press

Copyright© 2019. The Great American Business and
Economics Press. GABBYPress.

All rights reserved under Title 17, U.S. Code, International
And Pan-American copyright conventions.

No part of this work may be reproduced or transmitted in
any form or by any means, electronic or mechanical,
including photocopying, scanning, recording or duplication
by any information storage or retrieval system without
prior written permission from the author(s) and
publisher(s), except for the inclusion of brief quotations
with attribution in a review or report. Requests for
reproductions or related information should be addressed
to the author c/o Great American Business & Economics
Press, 620 Kingfisher Lane SW, Sunset Beach, N. C.
28468.

Printed in the United States of America
March, 2019.

ISBN 978-1-5136-4396-0

ISBN: 978-1-5136-4396-0

Table of Contents

Introduction: Christensen's Theory of Disrupting and Sustaining Technological Change.

Part of the theory of technology evolution in this book builds upon work done by Clayton Christensen, a professor of business at Harvard University.

"Sustaining innovations," states Christensen, "are what move companies along established improvement trajectories. They are improvements to existing products on dimensions historically valued by customers."

Sustaining innovations are primarily *technological innovations*, broadly defined as those that introduce a different set of features, performance, and price attributes relative to existing products and technologies.

In other words, sustaining innovations make an existing product more user-friendly.

The term "disruptive technology" is seen by Christensen from the perspective of senior managers in multi-national corporations.

For those managers, a disruptive technology, whether it originates in their own corporate organization, or from some outside organization, tends to disrupt the existing market demand of their existing products.

In Seeing What's Next: Using Theories of Innovation to Predict Industry Change, Christensen argues that it is possible to predict which companies will win and which will lose in a specific situation.

Christensen's thesis is that large, global companies blow it precisely because they do everything right.

He explains why the large companies that had listened hard to customers and invested like crazy in new technologies still lost their market leadership when confronted with disruptive changes in technology and market structure.

"After a radical disruptive technology takes root in new markets," said Christensen, "and after new growth is created, disruption can invade the established market and destroy its leading firms."

As Christensen notes, "The techniques that worked so extraordinarily well when applied to sustaining technologies, however, clearly failed badly when applied to markets or applications that did not yet exist."

Christensen called this problem the *innovator's dilemma*.

As applied to private corporations, he argued that market leaders have difficulty diverting resources from the development of sustaining innovations, which address known customer needs in established markets, to the development of disruptive innovations, which often under perform established products in mainstream markets but offer benefits that potential new customers would value over existing products.

The dilemma for private corporations is that the status quo arrangement with existing customers makes it difficult to give up on them in order to focus on new customers that do not exist yet.

Diverting resources from existing products is a risky proposition because there is no assurance that the future market opportunities will open up.

However, not diverting resources is also a risky decision because new radical, disruptive products cut into the low end of the current marketplace and eventually evolve to the point where the new products displace high-end competitors and their reigning technologies.

The biggest managerial incompetence in large corporations, according to Christensen, comes from a mental allegiance to the status quo.

In much of his work, he explains that large corporations and organizations customarily develop mind-sets and processes that revolve around doing what they already know.

Once that pattern becomes established, managers have great difficulty justifying to others, or even themselves, the need to turn their processes upside down to respond to a barely emergent market change.

Of the 172 companies studied by Christensen, 95% reached a point where their growth in sales stalled to rates at or below the rate of growth in the GNP. After the period of time that the growth rate stalled, "only 4% were able to successfully reignite their growth even to a rate of 1% above GNP growth."

Christensen shows how the stock market price declined for companies whose growth rates stalled. In some cases, the decline in stock price represented 75% of the previous stock price.

He suggests that the evidence of the pending death of the company appears in the form of declining stock prices.

"Probably the most daunting challenge in delivering sales revenue growth," states Christensen, "is that if you once fail to deliver it, the odds that you will ever be able to deliver it in the future are very low."

Consistent declining top line revenues causes the stock price to decline.

In other words, there is a type of ratchet-down effect in revenues that does not allow the company to recover. The biological description of this effect is called "extinction."

Christensen makes an important distinction between the technological features of the new product and the buying habits of consumers. The functions and features of the new technology product are what constitute the "market" for the new good, not the buying habits of the users.

"It is the circumstances in which customers find themselves the features are the critical units of analysis," writes Christensen, "not the customer."

Christensen uses the example of how workers on the morning commute to their jobs view the possible range of goods that fit into their mental image of what economists have called their utility function.

He notes "...In the customer's minds, the morning milkshake competes against boredom, bagels...etc."

The significance of Christensen's insight about technological features is that users who see the new product for the first time are engaged in a mental imagination process of trying out mentally how the new product may fit into their utility function.

If those potential consumers decide to buy the new product, they may contribute to the creation of an entirely new future market.

The macro economic effect of the innovator's dilemma is that when the executives choose the status quo, the result is a low rate of national economic growth.

The corporate executives described by Christensen generally do not imagine a future market.

Rather, the senior executives adjust production to the prices and profits to the status quo market.

However, economic growth comes from the new flows of profit and incomes in new future markets. The new future markets are created by radical new technology products.

Chapter 1.
An Overview of Theory of Technology Evolution.

The underlying fundamental human behavior being investigated in technology evolution is how the human brain processes information in the presence of a novel event.

Predicting the direction of technology depends on predicting what humans are going to do when they first see a new, radical product.

The brain, in this case, is aiming at individual sovereignty, or as biologists may interpret the behavior, individual control over the environment.

There are two different brain processes that are relevant to the theory of technology evolution. One process is in the brain of the owner of the firm, the first time she sees a new product or new production technique.

The other brain process is the consumer, the first time she sees a radical new product for the first time, and tries to imagine how that product may fit into her welfare function.

The brain is acting, in a biological way, as the "searching/ selection" mechanism for the product mutation/innovation process.

Products currently selected in the existing market are more "fit" than those not selected. In order to separate the various processes under investigation, Melanie Mitchell, in An Introduction to Genetic Algorithms, breaks the two processes into a "search space," and a "selection space."

The search space refers to some collection of candidate solutions to a problem and some notion of distance between candidate solutions.

In the science of biology, her search space is "the collection of all possible (protein) sequences. As applied to economics, her search space would be a the economic technological possibilities frontier.

The "fitness landscape," a concept derived from Sewell Wright's work in 1931, is a representation of the space of all possible genotypes along with their fitness.

As applied to technology evolution, the fitness landscape would be a representation of market characteristics, such as income classes, along with the technological characteristics of the product, which satisfy consumer preferences.

Fitness and selection are two independent processes, with fitness being an outcome caused by adaptation and heritable selection within an existing seven-year period of market demand.

Fitness, as an outcome of both the process of adaptation (existing market) and selection, (future markets) exhibits "pre-selective bias," in the sense that market demand in one period conditions market demand in the next period of time.

John Holland, in Hidden Order: How Adaptation Builds Complexity, (1995), explains what happens when owners of firms confront novel conditions and non-ordinary configurations of firms.

In the case of new conditions, Holland suggests that the owner of the firm "...combines tested rules to describe novel situations."

Holland describes how the brain decomposes the new situation into familiar parts in order to apply rules of decisions in the past that had been successful in similar situations.

The brain is sorting and shuffling thousands of mental images, searching for decisions about the solution to the new situation. "When one hypothesis fails, notes Holland, the brain comes up with "competing rules that are waiting in the wings to be tried."

The brain of the consumer goes through a similar sorting and shuffling process.

As applied to economic evolution, new products (phenotypes) within an existing market compete with other products with similar technological characteristics (genotypes). The brain of the consumer most likely had seen a prior version of the new product, and applies that experience to her product selection rule.

In terms of adaptation (product competition), successful products are selected by the consumer, (mutate) and pass their successful genes to the next generation of products (mutation during redundancy).

In the case of sustaining innovation, the mutation process can be described as asexual, directed, and positive, yet the future economic outcome of the selection process is contingent.

Applying the biological metaphor to technology, any population of products with the properties of multiplication, variation, and heredity, will evolve in such a way that the component entities will acquire characteristics ensuring their own survival and reproduction.

However, during the time that the existing products are surviving and multiplying, they are on their way to extinction, without the introduction of new genetic material, obtained from two-parent genetic crossover.

Successful demes of products in economics are linked to existing markets, and the processes of adaptation/competition are occurring in markets with declining marginal profits.

The demes themselves are evolving, and the market demand characteristics are also evolving as consumers gradually shift buying preferences towards more technologically advanced products.

As Fred Hoyle notes, in Mathematics of Evolution, (1987), "Once the range of improvements conferrable by single base-pair changes have become exhausted, a species cannot evolve further... because the range of genetic adaptation has become exhausted."

Some humans, like the corporate executives described by Christensen, tend to resist technological change, and try to use social and political mechanisms to retard technological change in order to hold on to their status quo profits.

Other humans embrace change, and see new opportunities in the future that they can capitalize on for their own benefit.

The appearance of something new, in this case a new technological product, causes humans to imagine the future possibilities that the new thing creates.

Macro technical change is caused when a great number of consumers start imagining the same future.

In other words, technical change is a social phenomenon caused by the interaction of humans engaged in social/business relationships in the current period of time.

Technological change is not the same phenomena as technological evolution. Technological evolution is explained when current rates of technical change create new future markets.

The future markets that consumers are imagining creates the conditions of contingency that the Bayesian analysis can investigate.

Given a set of prior expectations about the future, the new product may cause a new set of expectations, which can provide probability distributions, or estimates, of the most likely future market that may emerge.

When consumers shift their buying preferences from old products to new products, their new buying habits create new flows of income that did not exist in the prior market.

Part of the new income in the future market is a result of increased productivity, meaning that output increases with reduced inputs in the production unit.

Part of the new income is in the form of profits related to the purchase of new goods, produced by new production units.

Another part of the future income is in the form of wages and salaries paid to people who work in the new firms that are producing the new products.

The new future markets created by technical product innovation represent an entirely different economic structure, with its own internal dynamic of growth, than the current economic environment.

The market selection process of consumers occurs when consumers first see the technologically-superior products.

The imagination in human brains of consumers is only remotely connected to changes in market prices in existing markets.

High rates of sustaining innovation in the current market may, or may not, be associated with radical technological innovation in products, as described by Christensen.

Pier Saviotti describes why the supply of entrepreneurs in a distinct economic region is so important to the process of technical change. (Technological Evolution, Variety and The Economy, 1996).

Generally, the entrepreneurs are drawn from the ranks and staff of existing large companies.

Entrepreneurs meet with each other in their social business networks, and discuss how to solve technical problems.

Saviotti writes,

> The knowledge of engineers, scientists, managers, technicians, etc., involved in the implementation of the technology becomes specialized around the process, technical and service characteristics used.

In other words, within the existing social-business network of skilled individuals, there is a shared specialized knowledge about production processes and markets.

This specialization in knowledge creates networks of communication and power relationships which reinforce the allegiance to the status quo.

One reason an economy develops specialized technological knowledge is related to the "...specific institutional configurations and by the cumulative, local, and specific character of the knowledge that the institutions possess."

Charles Kindleberger, in World Economic Primacy, (1996), noted how a certain set of cultural values tended to favor an attitude towards technical innovation.

He characterized this attitude as the

> ...capability and will of individuals, companies and governments to break free of existing habits, perceptions, institutions, and task allocations, in order to revise them in light of constantly changing circumstances and developments.

Often, the solutions to technical problems involve the creation of new products or on-the-factory-floor production techniques that cannot be created within the existing older firms.

Entrepreneurs with a certain personality may become frustrated that they cannot solve the problem within the existing firm.

Those type of entrepreneurs either stay with their existing firm to attempt to create the new product, or they leave the old production units to create new ventures.

The new ventures that they create are based upon the knowledge they gained about how things work in the old firm, and with their ideas about how to make the new venture more productive than the older units.

The entrepreneur is the agent of technical change that links the unknown future of consumer preferences to market possibilities that result from radical product innovation.

Entrepreneurs perform the economic function of creating the future markets by imagining how that market will work. They provide the guesses of prices and profits, and how technological change in production units will interact with, as yet unseen, consumer preferences.

In other words, consumers imagine how the radical new product may fit into their utility function, and entrepreneurs imagine how that future market may function.

According to McAdams in Paths of Fire, (1996), the entrepreneurs have a "creative vision" in their capacity to "anticipate a new convergence of consumer preferences and technological possibilities."

Peter Temin makes this same point in his article, "Entrepreneurs and Managers." (1991), He states that

> ...entrepreneurs are the agents of change, ...(they) see new opportunities, invent new machines, discover new markets, ...(they) perform a different function from that of the manager, who works within a known technology, organization, and market.

Sarah J. Marsh and Gregory N. Stock, in their article, Creating Dynamic Capability: The Role of Intertemporal Integration, Knowledge Retention, and Interpretation, (2006),state:

> Uncertainty about the ability for technological knowledge to be transformed to meet market demands, lack of complementary technologies, the lack of developed markets for a given technical feature, and other types of uncertainty add significant challenges to organizations as they develop products for future markets.

As Marsh and Stock point out,

> Suppliers, customers, consultants, and the results of benchmarking and reverse engineering all provide sources of external knowledge that may be utilized in the new product development process, as do patent applications, scientific and trade publications, and conferences.

Part of the economic effect of technological change is that the increased production efficiency in the new ventures tend to make the older units obsolete and unable to compete.

In making the distinction between unknown costs and risks of the future, and the known costs and risks in the existing old production unit, Temin and McAdams hit upon the single greatest economic contribution that entrepreneurs make to technological evolution.

The reason that the technological skills of engineers and scientists is so important is that "technology" is defined as a body of knowledge about how things work.

The short-term competition between firms is based upon prices for existing products.

The primary form of long-term market competition between existing firms and new ventures is based upon competition for technological knowledge.

Understanding where technology knowledge is created, and then watching how new knowledge diffuses in an economy, provides the basis to understand the direction of technology.

In "Technical Change as Cultural Evolution," (1993), Richard Nelson states that

> ...technology needs to be understood as consisting of both a set of specific designs and practices, and a generic knowledge that surrounds these and provides an understanding of how things work, the key variables affecting performance, the nature of currently binding constraints and promising approaches to pushing these constraints back.

Nelson identifies the two key components of technology as the learning component and the knowledge component. Learning about technology generally occurs as a result of social processes characterized by *learning by doing*.

Learning by using, and *inter-industry learning*, results when suppliers, vendors, and customers share knowledge about technology.

Each case of learning occurs within a network of social-business relationships, some of which occur on the floor of the production unit working with machines.

Saviotti (1996), notes that innovative firms,

> ...tend to cluster in those (areas) that were already innovating countries...this specificity cannot be explained by factor endowments, but is more likely to be caused by specific institutional configurations, and by the cumulative, local and specific character of the knowledge that the institutions possess.

The accumulation of technological knowledge, and the pace of technical change, are contingent outcomes of the social and political institutional structure of a nation.

In his historical review of technology, Robert McAdams (1996), noted that historians share a widely-held belief about the origins of venture capital in the eighteenth century that fueled the industrial revolution.

McAdams wrote,

> ...all of the sources are in agreement that considerable increases in disposable wealth came into the hands of a substantial elite during the later part of the eighteenth century. Perhaps it is not so much their diversified and growing desires as consumers that quickened the pace of technological advance (during that era) but rather the increasing supply of potential venture capital for which this elite was beginning more aggressively to seek new avenues for profitable investment.

McAdams highlights two important factors about the origin and disposition of venture capital.

First, he describes venture capital in the hands of the financial and political elite as "disposable wealth," not disposable income. In the case of venture capitalists, the general source of wealth is the capital gain they achieve on the sale of their ventures.

Second, McAdams noted that the elites in the eighteenth century were looking for new ventures to invest their wealth. The elites were venture capitalists, not commercial bankers, but sometimes in economic literature, the role of venture capitalist is incorrectly used interchangeably with the role of commercial banker.

If, and when, the venture capitalists decide to invest in new ventures, they contribute to technical change.

The outcome of investing in new ventures is not the same economically, as when commercial bankers make loans to existing companies.

Bruno and Tyebjee (1982), identified the supply of venture capital as one of the top three factors that seemed to explain where entrepreneurs came from before the became entrepreneurs.

The supply or absence of venture capital in a region is a matter of historical contingency. The supply of venture capital in the current period of time depends on what the venture capitalists did with their capital gains in the previous period of time.

If a new pool of venture capital is created, and if that supply of potential new disposable wealth is then re-invested into subsequent new products, equipment and machinery, then the conditions for the emergence of a new future market.

The pool of capital that becomes available for reinvestment does not depend on cash flow from old production units, nor is it related to the conventional loans made by commercial bankers.

The potential pool of capital is created by profits on exit events made by the venture capitalists.

The reinvestment decision depends on what the venture capitalists in the region decide to do with the disposable wealth that is created by the forces of technical change.

If the economy is dynamic in the creation of new investment opportunities, and if the venture capitalists decide to make a second generation investment in new technology, then there is a chance that a second round of new innovative products may be created.

If consumers shift their buying preferences from old, existing products to new technology products, there is a possibility that a new future market may emerge.

If that potential new future market is created, it is possible for technology to evolve.

Given a certain type of prior social and political conditions in political power and cultural values, the economy could be predicted to take a certain type of trajectory.

Given another constellation of political power, the economy would maintain the status quo.

In terms of evolutionary logic, the social and political conditions represent the "environment" within which the units of analysis adapt or modify their behavior.

The analytical framework for economic evolution is based upon an analysis of how the social structure affects the economic structure as captured by coefficients in input-output accounting models as they change over time.

The existing environment may be characterized as having social assets.

The stock of assets in a social structure, such as specialized knowledge in production, can be measured at different points in time. Economists sometimes refer to this idea as the initial factor endowments that make one nation's economy different than another nation.

In other words, the underlying premise of technology evolution is that, as the conditions in the social environment change, new opportunities for genetic technological combinations of radical new products, which can be tried out, or "selected" by the imaginations of consumers.

The theory of technological evolution depends on what the human brain imagines, the first time it sees a new product.

The brain, in this case, is aiming at maximizing the welfare of the consumer by imagining how the new product fits into the consumer's utility function.

The consumer has a risk that the new thing may not work as imagined.

Consumer risk retards the buying decisions of new products, thus contributing to the economic status quo and a declining rate of economic growth, as existing products become extinct.

Consumers shape the evolutionary path of the future market in a LaMarkian process that occurs in product selection, in the final demand market for finished goods.

The product is the same thing as the phenotype that evolves in Darwin's theory.

In the classification scheme of technology evolution, products are categorized by firms who share certain characteristics in common. Firms are born, firms grow, and firms die.

Some firms do extremely well in a short period of evolutionary time, and the owners are rewarded by high rates of profits.

If, during this period of high profits, these successful firms procreate new products, which produce new streams of revenue, the firms may continue to enjoy success in future periods of time.

In the theory of technology evolution, the product passes on technological characteristics that it has acquired through production.

Consumer market final demand is providing the selection force, and, depending on what the consumer does the first time she sees a new product, the final demand forces can have both positive and negative feedback mechanisms to both firms and products that tend to "direct" the future contingent pathways of technological evolution.

Chapter 2.
The Scope of the Metaphor Between Biological Evolution and Technology Evolution.

In the application of the metaphor, certain rules about human nature, the functioning of the human brain, rules of genetics, and rules of civil society, are called upon to explain how technology evolves.

The main questions of the new theory would be:

- How do new radical products originate, and
- How do new markets emerge from the introduction of radical new-to-the-world products (phenotypes) in an economy.

This application of the biological evolution metaphor follows the 1962 suggestion by Max Black, that scientific metaphors, "...cause a profound reconception of the subject matter."

In this case, the re-conceptualization of technology evolution is from an understanding of economics as a Platonic ideal system that always seeks "equilibrium," to an understanding of economics as a dynamic evolutionary system.

The power of the re-conceptualization is that it allows the new theory of technology evolution to accommodate a range of facts and observations about the real world that the prior equilibrium theory could not address.

The two processes in technology evolution are the technological changes in products, and the evolution of consumer demand for finished goods, in the final demand market.

And, for long durations of time, the relationship between the two processes is static, generally called equilibrium.

Equilibrium theory is useful for explaining the existing economic environment, especially as it relates to the concept of equilibrium during a fixed or certain time period.

Darwin's theory is most useful for describing the processes involved with an economy breaking away from equilibrium, either on an upward trajectory of growth, or on a downward trajectory of economic decline.

Max Black (1962), said that the metaphor works by applying to the principal subject, in this case, technology evolution, a system of "associated implications," or characteristics, of the subsidiary, or source theory, in this case, biological evolution.

Biological theory, prior to Darwin, was framed by the philosophy of Platonic idealism, as the source theory.

Prior to Darwin, the notion of the biological "ideal" was the standard to which real world biological events were compared.

The failure of real world biological events to conform to the ideal was taken by scientists as a "measure of imperfection of nature."

After Darwin, biological evolutionary theory could be reconceptualized in a way that allowed it to accommodate new ideas, like molecular genetics, and new scientific methods and processes like proteomics.

In the metaphor of Darwin's theory, (On the Origin of Species by Means of Natural Selection, or the Preservation of Favoured Races in the Struggle for Life, 1869), the analogy to Darwin's phenotype in technology evolution is the product, which is produced by a firm, and its owners, who make decisions about the firm.

Among individual populations of different species, Darwin found enough diversity, and enough blind variation, to sustain an evolutionarily adaptive selection process.

In his theory, there are three major objects of inquiry related to biological evolution.

First, entities within the population must multiply and reproduce, based upon some observable, logical, and natural scientific explanation, in his case heredity.

Second, the entities must be subject to categorization by how they vary among themselves and between species.

Third, the phenotypes are subjected to a selection process associated with the unique environment, and, over time, certain species adapt to the demands of the environment, and pass the selected traits, through heredity, to their offspring.

Darwin's theory of natural selection is not a perfect fit for explaining the evolution of technology.

In contrast to Darwin, in technology evolution, the selection of technological variety is not a randomly occurring event, it is a reasoned and strategic outcome that is based upon how human brains imagine the future, when they see a new product.

In Darwin, the selection process is caused by species adapting to their environment, with the most fit species surviving to the next generation.

The application of the metaphor raises the question: How is technology evolution like biological evolution?

The mechanisms of genetic processes, at both the molecular and phenotypic level, are used to explain the mechanisms of technology evolution.

In other words, Darwin's theory can accommodate two important economic concepts about how the future economic structure may evolve, given a set of antecedent conditions.

In the application of the biological metaphor to technology, biological evolution provides a means of transferring inferences from the source theory of biology to inferences for the target theory.

New products that inherit technology from old products, (sustaining innovation), do so through technology heredity and mutation, which leads to greater rates of market selection, because consumers had seen something like the new product before.

In the logical chronology of the passage of time, after technological variety has been created, the selection process in the marketplace by consumers is not purely adaptive.

In other words, new products, which look and function like old or existing products, have a greater rate of market adoption by consumers, when they first see the new product.

On the other hand, the more that new products look like old products, the less likely the introduction of new products will lead to new future market emergence.

The paradoxical result is that new radical products with new technology, obtained through two-parent genetic crossover, have a lower initial rate of market selection, but have the greatest effect on new future market emergence.

Consumers who had not seen the "new-to-the-world" technology are reluctant to buy a radical product that may not work the way that they expect.

The basic concept being described is, that in a market characterized by asexual adaptive behavior, the most recent technological version of the product displaces the parent product.

Asexual product technology evolution is similar to Christensen's concept of sustaining innovation.

The scope of the metaphor is described in Diagram 1.

Diagram 1. Scope of the Metaphor Between Biological Evolution and Technological Evolution

Biology Units of Analysis Environment	Basic Rule Algorithms	Antecedent Conditions	Existing Environment	Future Environment
Genetic molecules DNA Demes Genotypes Phenotypes	Units must multiply Units must vary Variation is heredity Adaptation Selection Survival of fittest	Initial population Birth rates/death rates Inbreeding	Current advantaged phenotypes Current rates of genetic evolution	Heterozygosity leads to new species VS. Muller's Ratchet of inbreeding

Time Chronology of Analysis

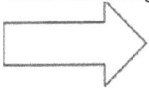

Technology Units of Analysis Environment	Basic Rules/ Algorithms	Antecedent Conditions	Existing Environment	Future Environment
Humans Products Firms Markets	Diversity Variation Selection Adaptation Obsolescence Civil rules Social rules Moral rules	Initial population History Culture Existing infrastructure Industrial clusters, skills Technological envelope	Birth/Death rates of firms Rate of technological product innovation Local attractor points Regional economic structure [A] Final demand market selection Intermediate demand market selection	Bifurcation growth points Bifurcation decline points Future economic structure Eigenvalue from [A] to [A']

The asexual, sustaining innovation is bought first by the highest income classes because the new product's price is higher than the earlier version, and it contains new technology that the highest income classes are most likely to have seen and used before.

The higher income classes have an income level that allows them to take the greatest risk in making a risky new purchase on a product that may turn out to be non-optimal.

As newer versions of the product enter the market, the earlier technological versions drift down in price and are bought by lower income classes. Even as the market widens, the contribution to net marginal profits from the earlier product version is declining.

In evolutionary technology theory, the basic premise is that some event or condition in a prior time period affects, or "maps on to" events in later time periods.

The basic causal mechanism of mapping from the past to the future is the genetic transfer of technical characteristics between generations of products.

The prior conditions set the environmental constraints within which the consumers adapt or modify their behavior by selecting products.

In other words, the antecedent economic conditions changed into the current conditions as a result of some factors or forces, and the purpose of evolutionary technology theory is to provide the explanation of why historical conditions changed to evolve into the current conditions.

There is a Bayesian probability attached to the antecedent configurations of political power and cultural values, based upon the degree of belief about the future held in the mental images of humans in a distinct economy.

The Bayesian statistical analysis would also be applied to beliefs and expectations in the current period that related to the appearance of entirely new technological coefficients in the economic interindustry matrix.

Then, based upon the application of rules and regularities from the biological metaphor, the future market structure is predicted, based upon the application of Bayesian statistical analysis of the prior beliefs about what humans do when they "see" the new product.

The general analytical terms used to describe the evolution of technology are:

- social structure,
- interindustry intermediate demand structure,
- information networks,
- industrial technology relationships between firms.

In Diagram 2, the logical theoretical framework of evolutionary economics describes how the prior social structure affects the current economic structure, which is then shown to be affecting the future market structure.

The logical linkage from the market to the social political structure is a result of competition and rent seeking behavior related to income distribution, not prices.

One of the important theoretical feedback loops, in Diagram 2, is between the average production techniques currently in use, and the "best practice" coefficients, when they first appear in the economy.

The relationship between the average a_{ij} and the best practice a_{ij} provides some estimates of probability distribution at future points in time of how the economy would evolve under both consumer behavior selection responses, and the behavior of the owners of firms.

It is the adaptive behavior, by owners of firms, who imitate the most recent best practice of their competitors, that causes productivity to improve.

The imitation of best practice in production is akin to passing along genetics in the production side of the product.

Passing along best practice genetics is like a production efficiency improvement, which has the effect of dropping production costs, which drops profits, while the market is saturated with lower priced goods produced by the best practices.

Diagram 2. The Sequence of Events Underlying Technological Evolution

Elapse of Time	Social/Political Setting	Structure of Production Relationships	Intermediate/ Final Demand Relationships
Antecedent Historical Conditions Of Production Technology	Institutional networks of knowledge creation. Initial distribution of income. Cultural values regarding property.	Regional input-output technical coefficients. Population size of region. Regional industrial clusters.	Historical rates of consumer selection of products. Social income classes. Market demand for regional products.
Direction of causation	Social/Political Environment	Production Transformation Matrix	Exchange Markets Equilibrium
Time Horizon Of Production Technology	Institutional networks of knowledge diffusion. Allegiance to status quo. Birth/Death rate of firms.	Addition or deletion of rows or columns in regional I-O transaction matrix. Rate of product or process technological obsolescence. Relationship between regional production and global production units.	Regional intermediate demand markets. Final demand markets. Labor markets. Debt capital markets. Equity capital markets.
Evolution of Technology	Social/Political Change/Stasis	Future Production Technical Coefficients Matrix	Micro Market Bifurcation
	Distribution of income. Distribution of political decision making. Allocation of capital surplus.	New firms with technologically different products in industrial clusters. Change in income/employment multipliers. Technological trajectory of regional eigenvalue.	Adaption/selection of consumers of new products. Venture capitalists selection of new ventures. Death rate of existing firms within regional industrial clusters.

One of the important theoretical feedback loops, in Diagram 2, is between the average production techniques currently in use, and the "best practice" coefficients, when they first appear in the economy.

The relationship between the average a_{ij} and the best practice a_{ij} provides some estimates of probability distribution at future points in time of how the economy would evolve under both consumer behavior selection responses, and the behavior of the owners of firms.

It is the adaptive behavior, by owners of firms, who imitate the most recent best practice of their competitors, that causes productivity to improve.

Paradoxically, increasing rates of technology productivity improvements that result from adaptive behavior of firms, hastens the arrival of zero profits for all firms with the "best" technology in production.

Productivity improvements in existing product markets do not necessarily lead to new markets, and it is new markets, with new income distributions, that cause economic growth.

The rate of new venture creation and new technology commercialization must be very high to overcome the natural rate of technological obsolescence that accompanies loss of genetic diversity, via mutation during redundancy.

Mutation during redundancy describes an evolutionary process that begins in a steady state, or initial condition, of a stable population of product species.

As firms adapt the best technology in production, mutation during redundancy is causing loss of technological diversity.

The technological variety in production, and in new product technology, must be "stirred up" by radical new products, to avoid technological obsolescence.

And, in order to understand and explain the emergence of new markets, the concepts of state transition functions must be linked to the concepts of genetic mutations and genetic crossover, the precursor activity to understanding bifurcations in future economic demand.

Mutations, recombinations, and two-parent crossover of technology, are the processes involved in understanding how the state transformation functions describe the scientific logic of how technology evolves.

Future economic growth is based upon the birth of new firms, who introduce technologically different products than existing products, and the production of the products creates new exchange relationships in the intermediate demand market.

If, as a matter of contingent evolutionary process, these new technologically different products create new income flows, and if those new income flows modify the initial distribution of income in the economy, then future economic growth may occur.

It is not automatic, however, that growth will occur, even if income distribution changes.

In other words, there is not a perfect fit between biology and economics because market competition and allegiance to the political status quo intervene in the technology mutation process, causing contingent outcomes, which arise outside the scope of technological mutation.

There is no analogy in biological evolution to the effect of social and cultural values in technological evolution.

Because of the allegiance to the status quo, there are no technological factors "ensuring" the product's survival and reproduction.

Unlike Darwin, the survival of the fittest in technology depends on an outside factor of market selection and social/political allegiance to the status quo.

In both biological evolution and economic evolution, the initial product population size affects subsequent generations of technology evolution.

Following the classification scheme for initial populations provided by Austin Hughes, in Adaptive Evolution of Genes and Genomes, (1999), the processes of technological evolution would initially be placed within an analytical framework that had five features:

- Initially, there is a population of firms and products. The population is either growing or declining from one generation to the next.
- There is non-random, or "directional" asexual mating between products within existing technological industrial clusters.
- Firms and products have overlapping generations, and subsequent generations of products eliminate earlier products via a process of technological obsolescence.
- There are adaptive mutations of existing products, as owners of firms tinker with the outward appearance and inward technological features of the product, within the product's limited life-cycle of technological utility.

The genotypes are the technological characteristics of products, and positive assortive mating between products that look alike or function alike technologically, are called industrial clusters.

The industrial clusters represent "groupings" of firms by product, and the industrial clusters can be represented in input output transactions tables that have been modified by factor analysis to show technological similarities among products.

Individual humans affect their environment through their imagination and selection of technological variety, and they are, in turn, affected by the environment in which they happen to be living by "seeing" what other humans do when the other humans first see the new technology.

Like the analogy of singing into speakers to make a new record, the direction of causation in the genetics of technological information, as it applies to economic evolution, is a one-way flow from human imagination of technology to market selection.

Consumers in a market cannot sing into the speakers of technological variety to produce the new products that they desire.

What actually happens economically depends on whether the new genetic/technological information flow is horizontal and free, in which case inter-species breeding may occur and new species may arise, setting off a new market bifurcation in the final demand markets.

Or, whether the information flow is vertical and controlled, by social and political factors, leading to technological status quo, whose consequence is economic decline.

Chapter 3.
Building The Theory of Technological Evolution Upon the Foundation of General Equilibrium Theory.

In terms of building the biological metaphor on equilibrium theory, the most recent product X starts out life at the top of the income class food chain, with high prices targeted to high-income classes, generally in very small market niches.

As product X evolves in terms of production improvements, (sustaining innovation), the price of good X drops, and it becomes available to lower income classes.

The market widens but the profits, per unit of good X, are declining.

Finally, as a dominant product design is reached across all global production locations, good X becomes so cheap to build and distribute that it enters the mass, standardized global market.

From an evolutionary perspective, the entrance of good X into the global mass market is the point in time that marks the end of the technology innovation life cycle for good X.

This entire sequence of events can be captured and described by neoclassical theory.

In conventional equilibrium theory, most of the attention is on how the economy adjusts to an outside perturbation to return to equilibrium.

The same analysis can be turned on its head to describe how an economy breaks away from the current equilibrium, to a distant future attractor point.

Breaking away from equilibrium applies the insight of the biological metaphor to allow a re-conceptualization of equilibrium. In this case, the relationship depicted is not simply quantity and price, but production output and technological genetic diversity.

The environmental conditions for breaking away from equilibrium depends both on the supply of product genetic diversity, caused by firms seeking new sources of technological food, and environmental chance, in the form of initial technological conditions that allow for inter-species technology product breeding.

Diagram 3 describes how an economy slowly approaches the hypothetical point of equilibrium. Prior to breaking away, the economy can be visualized as slowly revolving around an equilibrium point.

Diagram 3. Economy Slowly Revolving Around a Hypothetical Equilibrium.

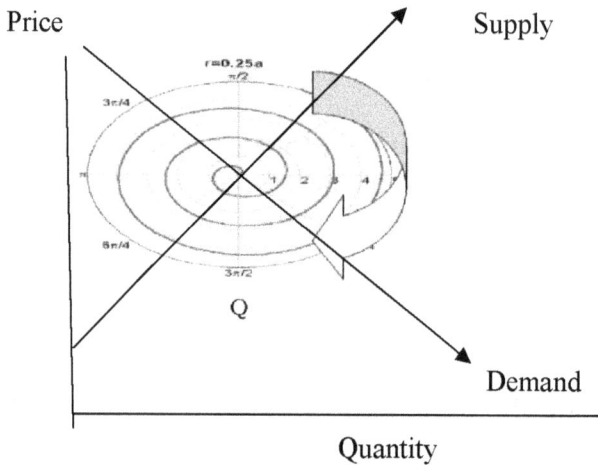

In the standard treatment of equilibrium, as the price goes up on the vertical axis, the quantity demanded decreases, along the horizontal axis.

The idea conforms to observed reality that buyers tend to buy less if the price goes up, and would probably buy more if the price goes down.

Also, as price goes up, sellers and suppliers would likely supply more, while if prices are lower, they would probably supply less, of whatever it is they are supplying.

The slowly revolving equilibrium point in Diagram 3 can be described as a relationship between production output and technology genetic diversity in time period one.

In Diagram 4, the initial supply of genetic technology diversity limits the gross output of production. As technology genetic diversity is used up in mutation through redundancy, (sustaining innovation), the economy approaches it theoretical equilibrium attractor point, in time period one.

Diagram 4. Economy Approaching Equilibrium Attractor Point In Time Period One.

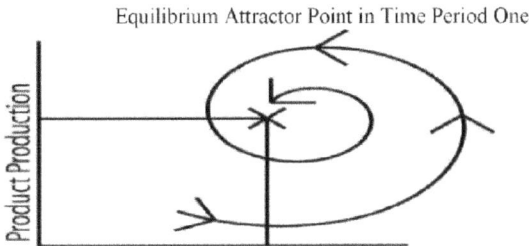

Equilibrium Attractor Point in Time Period One

From that equilibrium point in time period one, the economy can follow either an upward trajectory of economic growth, to a hypothetical attractor point in time period two, or, in the absence of a new supply of technological diversity, the economy can follow a downward trajectory of economic decline.

Diagram 5 describes what happens in the economy if product genetic diversity increases through genetic 2-parent product cross breeding.

Diagram 5. Economy Breaking Away From Equilibrium In Time Period One to Potential Attractor Point in Time Period Two.

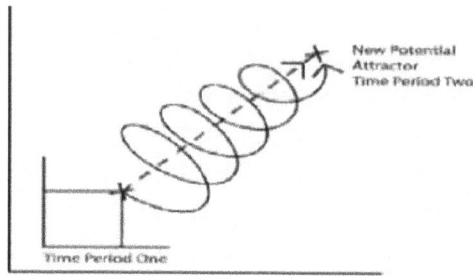

On the upward spiral of growth, the center points of each oscillation could be determined and connected as a line through those center points.

Even though the amplitude of the oscillations are declining, the center points would be tracing out a logarithmic curve whose rate of change would be approaching zero at some future point in time.

As the economy breaks away from the old equilibrium, each initial increment of growth adds a large amount of transactions to the cycle of growth.

Later increments of growth add to the cycle, but at a diminishing marginal rate.

The hypothesized future attractor point is where the marginal rate of change is close to zero, which represents the economy's next equilibrium point.

The economy reaches its future attractor point through a series of micro bifurcations in growth.

At some point in time, given new products and new investments, these micro bifurcations lead the economy to the potential attractor point in time period two.

The transformation from one attractor point in time period one to another attractor point would be described as a macro economic Hopf bifurcation.

The economy would have passed through an evolutionary Hopf bifurcation point where a new market had emerged. At some future point in time, after several micro bifurcations, the economic structure would no longer look the earlier economic structure.

The process of breaking away from equilibrium in time period one can also be described as a downward trajectory to a future attractor point of economic decline, characterized as Muller's Ratchet.

On the downward spiral of economic decline, the amplitudes of each cycle become smaller, representing the idea that the overall economic activity of exchange transactions is diminishing.

The rate of growth in the economy is diminishing because the product genetic diversity is being used up in mutation through redundancy, without the addition of a new supply of genetic material.

Once an economy loses the ability to generate new genetic diversity, it loses its neuronal pathways. If no new genetic material is added, the economy slowly drifts down through micro bifurcation points, called Muller's Ratchet of Economic Decline, through which a return to economic growth is very difficult because technological diversity is eroding over time.

Losing technological genetic diversity for an economy is just like the human brain losing neuronal pathways. Once the neuronal pathways are destroyed, the brain loses its ability to imagine new futures.

The application of the biological metaphor for this part of technological evolution is described by Austin Hughes as "mutation during redundancy." (Adaptive Evolution of Genes and Genomes, 1999).

As he applies the term, mutation is an inheritable change in the genetic material. In the case of products, a mutation would be an inheritable change in the technological features of the product, acquired from the parent, in an asexual process.

Asexual product mutation is adaptive and directional within an existing phenotype market environment, called the technological possibilities frontier.

The repertoire of genes within an existing technological possibilities frontier is based upon prior flows of information.

Some of the genes are expressed in the current product/current market environment, while other genes are latent, but able to replace or substitute for current genes, given the right set of environmental or information conditions.

Mutation during redundancy serves two functions in technology evolution.

The non-expressed genes serve as the supply of potential genes in the adaptive evolution of products, and also serve to set up a potential stage for a new market.

The consumers in the existing market "call forth" the non-expressed gene as the forces of adaptive competition change in an existing period of time.

As the genes are called forth, they are expressed in the form of product mutation that is asexual, adaptive and directional.

The genes called forth and selected by consumers implies that genes are inherited in the next generation of products, but once called forth, due to the one-way direction of information in genetics (singing into the speakers), they contribute to the sweep towards genetic fixation.

Fixation, used in this context, is analogous to equilibrium, as it is used in economics.

The radical technology mutations that disrupt the existing environment are least likely to be selected, and passed on, genetically.

The greater the rate of mutation that disrupts the existing environment, the less likely the rate of selection, but if selected, the greater the rate of future market mutation and bifurcation.

The loss of product genetic diversity is a result of sustaining technological innovation in the absence of two-parent technology cross breeding.

Diagram 6 describes the rate of change in economic activity as a curved line that connects the midpoints of each oscillation.

At the future Nash Equilibrium, the economy can spin around and around the equilibrium, for decades at a time. The economy has ratcheted down to a low level of macro aggregate demand, and has adjusted to the lower level of production.

Diagram 6. Economy Breaking Away From Equilibrium In Time Period One to Potential Attractor Point of Economic Decline in Time Period Two.

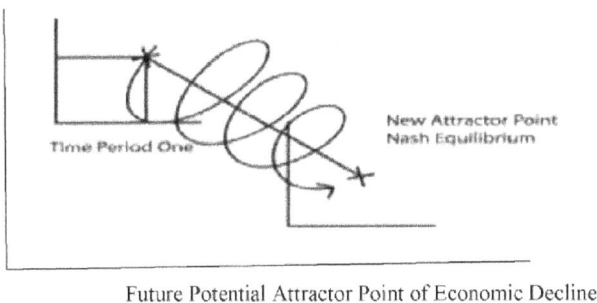

Future Potential Attractor Point of Economic Decline

The most recent real-world example of the Nash Equilibrium is the U. S. economy during the eight years of the Obama administration. The 2% economic growth rates during that period describe what the economy looks like at the lower level of technological diversity.

Diagram 7 describes an economy in equilibrium at the lower level of economic growth, stuck in the Nash Equilibrium, and unable to break free unless some event stirs up genetic diversity.

The economic pathway to time period two depends on the consumer selection in time period one. In other words, time period one is the antecedent environment, from which future technology evolution occurs.

If product genetic diversity increases through two-parent cross breeding, (radical innovation), there is a possibility that the economy may take an upward trajectory of economic growth to a future distant attractor point.

If genetic diversity does not increase in time period one because the sustaining innovation does not add genetic diversity, then the economy will follow a downward trajectory to a future Nash Equilibrium.

Diagram 7. Three Dimensional Representation of An Economy In Decline Approaching Nash Equilibrium in Time Period Two.

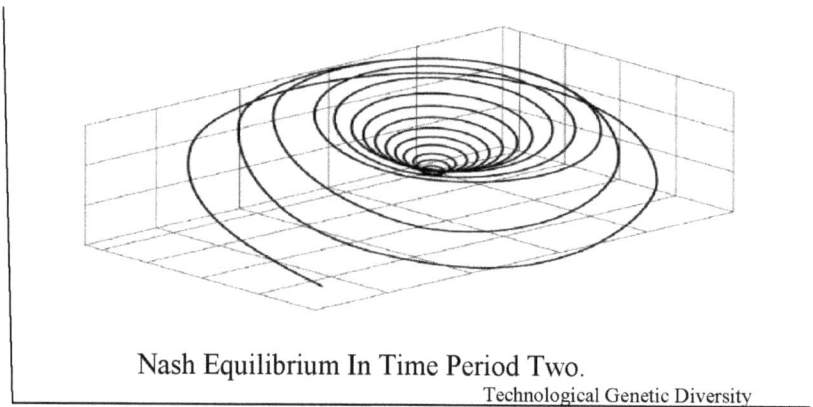

Nash Equilibrium In Time Period Two.

Technological Genetic Diversity

Diagram 8 depicts the two possible pathways an economy can take from the equilibrium point in time period one.

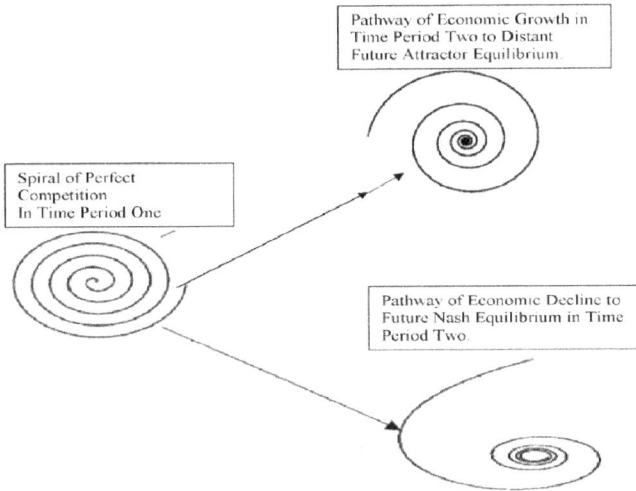

Christian Forst, a professor of genetics, describes the three evolutionary processes involved with understanding how an economy can break away from equilibrium. (Evolution of Metabolisms, 1999).

First, within the product sequence space, technological genotypes must be mapped in both the asexual and sexual heredity space.

Second, in the product phenotype space, technological selection must be mapped from product to market selection.

Third, in the geographical space, the topology of the technological intersections between industrial clusters within the region must be mapped, in order to understand how technological genotypes in one cluster end up in phenotypes in a new cluster.

The first process is characterized by a one-way flow of genetic information.

The second process contingently involves a two way flow of horizontal or vertical information, which affects the third process, which itself can feedback, via political manipulation of the environment, and affect the first flow of information.

The second process of mapping the technological information to the market selection process is based upon what consumers do when they first see a new technological innovation.

Great new technological products could be created in time period one, but if consumers reject the product, nothing happens in the economy to move it to a distant future attractor point.

In the application of the biological metaphor, consumers perform the selection process described in Darwin.

After consumers have made their selection, the economy adapts to its new equilibrium in time period two.

Diagram 9 describes the effect of consumer selection in time period one on the creation of genetic diversity.

The three processes explained in Diagram 9 are roughly the same as described by Forst. The first is consumers searching for a fit in product $x1$.

The second is owners of firms searching for consumers to buy $x1$.

If high income consumers buy generation $x1$, the price of $x1$ declines, allowing the lower income classes to begin buying $x1$.

As the price drops, firm owners attempt to add sustaining innovations to $x2$, which comes out at a higher price. High income consumers had seen something like $x2$ before and may buy it at the higher price.

At some point between product generation 1 and generation 3, technological genotypes in a different technology cluster may end up in a new product phenotypes in $x3$.

That new product had not been seen before by consumers, and the consumers have a risk that the new product may not "fit" into their welfare function.

The three biological process described by Forst are depicted in Diagram 9 as the interaction between consumer search for product fitness and producers search for technological features that consumers will select.

From a biological point of view, the chance environmental conditions in time period one, coupled with genetic competitive pressure in all three processes in time period one, create the conditions for breaking away from equilibrium.

Diagram 9. Relationship Between Consumer Market Genetic Selection Over Two Generations of Products, Product x1 and Second Generation Product, x12.

Dual Market Search First Generation Occurs **In Product Technology**	Market Genetic Selection First Generation **Product Fitness Operator**	Genetic Search Second Generation **Product Technology Fitness Landscape**
By Consumers Searching for Product Fit	By Producers Searching For Consumers	Price of first generation product drops
Sorted by Income Class	Sorted by Vintage of Product Technology	Prices	Customers

C_1 X_1 G_1 X highest → Sales X_1 X_{12} second generation

C_2 X_2 G_2 X Sales X_2 X_{22} second generation

C_3 X_3 G_3 X lowest Sales X_3 X_{32} second generation

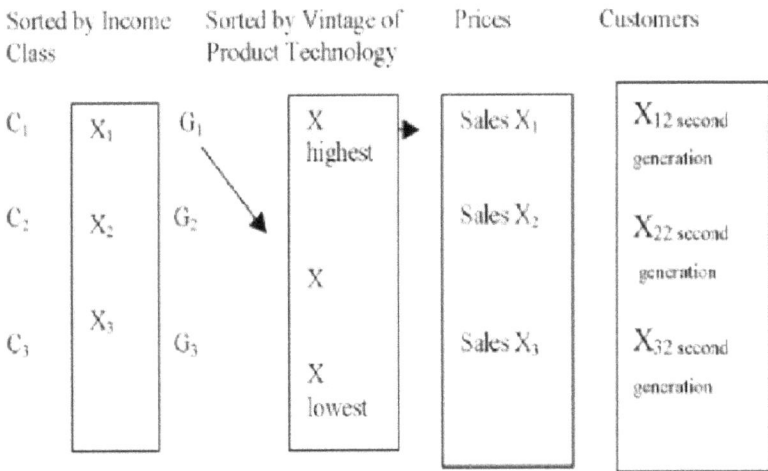

C = Income Class
G = Product Generation

As described by Forst, breaking away from equilibrium would be rare, and dependent on just the right mix of events to push the technological evolutionary process to the strange attractor point that looked entirely different than the current state of equilibrium in time period one.

The probability of a two-parent technology crossover leading to a bifurcation in market demand for a new product is contingent upon what consumers do when they first see the new product.

Under conditions where a product crossover in technology between two species occurs, if consumers select that product, the probability for market bifurcation exists, where demand characteristics between the two market environments change.

Unlike the selection process in Darwin, the consumer selection process is influenced by political and financial interests who are committed to keeping $x1$ alive, as long as possible.

Their allegiance to the status quo retards radical technological innovation in time period one, consequently retarding macro economic aggregate demand.

Existing firms producing $x1$, and existing commercial bankers who may have loaned money to those firms, have a motivation to control the direction of technology in order to maintain their incomes.

"Research and development is directed towards shaping and refining knowledge in very specific ways," states Peter Hall.

The shaping of knowledge that Hall is describing is an allegiance to the sustaining innovation in $x1$. (Innovation, Economics and Evolution: Theoretical Perspectives on Changing Technology in Economic Systems, 1994).

Hall explains that, "Research and development is worth doing only if it generates a product or process of commercial value...which both fits with the firm's existing capabilities and meets market requirements (of existing consumers)."

There are two adverse environmental fluctuations at work in any economy in time period one.

First, existing products, in existing markets are experiencing their own rate of technological obsolescence, as a result of the asexual process of technological evolution, as the repertoire of non-expressed genes in the technological possibilities frontier are used up, as they are "called forth."

Second, in the final demand marketplace, the rate of product obsolescence is hastened by competitive price pressures which are dropping the marginal rate of profits to zero.
The market entry of new firms producing $x1$ leads to an outward shift in the supply of good $x1$, creating a surplus of production for good $x1$.

As the supply curve shifts outward, the price of good $x1$ goes down. As the price goes down, the profits associated with good $x1$ for the initial firm are eroded.

This process of price reduction and profit elimination continues until equilibrium is established between supply and demand at a much lower level of aggregate economic demand.

For the initial representative firm, the new equilibrium for the level of output will be where the price for good $x1$ equals the marginal cost for the firm of producing good $x1$, which now becomes the minimum value of the average total cost to produce good $x1$.

Existing firms, with fixed production on the plant floor and fixed inputs, are on a pathway to extinction because, in perfect competition, new firms, with new production equipment, and less labor used in production, produce the good more cheaply and sell at a cheaper price, a sequence of events which continually eats the profits of the existing, incumbent firm.

At the moment that the new firm enters the market, their production technology to produce good $x1$ represents the very best, least cost method of production.

The new firm's production technology becomes the new competitor to beat, in terms of costs, and acts as an attractor, or reference point, for all other firms that are adapting their own production to the new environment.

In order to compete with the best practice production technology, managers of firms with average technology may choose to cut variable costs, the most likely candidate being labor costs.

Other firms, with older production technology may not be able to catch up, technologically, with the new best practice.

The senior management in those firms may choose to cut costs down to average total costs, so that they can at least cover their variable costs as they cope with their declining sales and declining profits.

From an evolutionary perspective, the competition is between a population of existing firms who produce $x1$, and potential new firms, not yet born, who would like to make the profits on $x1$, but to do so, must be born and then enter the market in order to take away the profits currently being made by the existing firm.

Products produced by mutation during redundancy, given a fixed level of market demand, reach a stable age distribution within the economy, without increasing technological diversity.

In the case of asexual haploid product evolution, new vintage production technology, without radically new technological features being added to the parent product technology, hastens the time to obsolescence because it speeds up the rate to zero profits.

As products become more standardized in their technological features and more uniform in their production process, their marginal profit is headed to zero, and as the profit heads to zero, the political manipulation of the rules becomes more pronounced as income competition regarding maintenance of the status quo intensifies.

Once the latent, non-expressed genes (technology features) in period one are "called-forth," for $x1$, they are used up. Having entered the mass, global market, x1 is on the pathway to technological obsolescence.

As a result of genetic technological inbreeding (sustaining innovation), within the technological envelope, the genetic heterozygosity within the economy is reduced.

The product species, x1, cluster around the parent technology, with potentially better mutations being killed by the political control of technological knowledge.

At the stage of stable product populations and stable firm populations within the economy, the economy is characterized, in time period one, as an environment of the "Living Dead."

The final blow to survival is the chance introduction of technology that eliminates the product from the environment and sets the economic trajectory on the pathway to extinction.

The firms and products may still be walking around and breathing, but they are soon going to die, without the addition of new genetic technological material.

Unlike the loss of genetic variation in Darwin, which is a random naturally occurring event, the product technological inheritance in time period one is not random but positive and purposeful, and based upon the "fitness" selection process made by consumers.

In economics, the "fitness" of the product phenotype is analogous to adaptation in biology, and adaptation in Darwin is the economic equivalent of market competition.

The product selection of consumers leads to or "causes" variation in technology to decline as a result of consumers selecting existing products because they do not have radical new products to choose, as an alternative.

They have no new products to choose because of political manipulation of the rules that have the effect of eliminating the rate of investment in new products.

As the competition over legacy income distribution in the dying economy intensifies during the economic meltdown to the Nash equilibrium, the institutional political structure of the region undergoes a profound transformation.

In the once, vibrant, open and diverse environment that previously had created and diffused knowledge, income had been growing and distributed according to merit.

The economic growth was based upon self-generated knowledge and technological innovation, which was commercialized in the form of new ventures.

The risk associated with creating the new ventures was diminished as a result of "trust" and the adherence to the rule of law.

The economy is headed to the Nash equilibrium, where political and financial elites are engaging in political rule manipulation to hang on to their legacy income, which destroys trust and adherence to the rule of law.

However, the uncertain evolutionary outcome in technology mimics the evolutionary processes in Darwin.

Both evolutionary outcomes are contingent, because nothing about the outcome of evolution is certain.

Other evolutionary outcomes could have easily occurred, if a different combinations of mutations had occurred in time period one.

What happened in the past economic history could easily have been something else, genetically, if certain types of genetic bifurcation and technology crossover combinations had occurred.

Chapter 4.
Sustaining Innovation as Asexual Reproduction.

The asexual product innovations usually show up in slight variations of technological characteristics of the parent product technology, and in slight improvements in the manufacturing process, as the multinational corporations seek out advantages associated with tax incentives and skills that complement the corporate core technology.

Most "mating" among members of the product phenotypes within an industrial cluster is "asexual," in the sense that the technology being mated is previously existed in the genotype technological envelope.

In an asexual population of biology of phenotypes, would be equivalent to product phenotypes in the case of product mutation during redundancy.

The process of recombination does not add technological (genetic) diversity into product technology, but tends to use up the potential technology within the technological possibilities frontier in a process called mutation during redundancy.

The asexual process is more like genetic recombination, where existing alleles are recombined and shuffled in the existing genetic pool, without adding new genetic combinations in a new genetic pool.

In any given year, an existing product may acquire a new outward appearance that consumers seem to prefer over the previous appearance.

A mutation via recombination with a very small technological advantage is more likely to be selected by consumers because buyers are more familiar with the appearance and use of the product, the first time they see it.

If the owners of firms determine that the new appearance is marginally profitable, the very next product incarnation may include this acquired characteristic, in addition to some new characteristic.

The consumer demand within the genotype is acting to select the most fit product, and owners of firms are continually tinkering with the outward appearance and inside technological features of products to marginally make them more attractive, given an existing level of market demand.

John Maynard Smith estimates that 55% of the asexual population dies selectively each generation. (Evolutionary Genetics, 1998). A generation in human evolution is about 20 years.

Applying the biological metaphor, this death for a product phenotype occurs when the product becomes obsolete. The time to obsolescence for most technology products is around 5 to 7 years.

As described by H. J. Muller, "...in each generation, there is a chance that, despite their high fitness, all (deleterious mutations) will die without leaving offspring. If so, the optimal class is lost, and can only be reconstituted by back mutation." ("The Relation of Recombination To Mutational Advance," 1964).

If no other new technology is incorporated in the existing product, then a slow rate of genetic mutation is occurring in the economy that eventually leads to a Nash Equilibrium.

In an asexual adaptive system, product heuristics mutate, in a process of sustaining product innovation, and consumers and capitalists modify their behavior by improving strategies in a constrained maximization environment by buying the new and better goods that are produced at lower costs than existing goods.

During the period of time to the Nash Equilibrium, marginal profits are declining, but the sustaining innovation products are sold at prices that yield a slightly higher (marginal) profit for a brief period of time.

In the case of recombination of technology, products within the product genotype "acquire" characteristics from existing product technology in order to better adapt to the environment.

In the case of asexual heredity, for example, Oster, Guckenheimen and Ipaktchi, offer a model for populations with overlapping generations. "If the population breeds continuously, so that generations overlap, then the appropriate model is ordinary differential equations." (The Dynamics of Density Dependent Population Models, 1977).

Oster et.al., continue with the application of their model to a Lotka-Volterra predator-prey population.

The Lotka-Volterra model is based upon food that is eaten in a defined geographical territory. As they point out, "virtually all of the models for predator-prey systems possess either a stable equilibrium or a stable limit cycle."

In their model, the limit cycle oscillations are directed inward due to the finite population limitations. They suggest that the predator, in the absence of prey, dies out exponentially, the hypothesis, which they test with their model.

In their case of overlapping population generations, with limits on food, they very reasonably predict that the population of predators would reach some stable equilibrium.

However, for certain types of initial conditions in the environment, for example an environment that contained 3 or more species, the possibility exists for higher order bifurcations. (radical innovation).

They state, "Successive bifurcations beyond the first occur when the eigenvalues of the Poincare map passes outside the unit circle, headed towards a contingent vague or strange attractor."

In the application of the biological metaphor to asexual technology evolution, the "food" being sought by firms in the product genotype is "knowledge."

In an asexual sustaining technology environment, the food eaten by firms is passed along to the next generation of product phenotypes as an "acquired product characteristic."

The evolutionary process being described in asexual evolution is Lamarckian inheritance, where the product phenotypes in time period one pass on their genetics to the next generation of slightly different product phenotypes.

The difference between Darwin and LaMarck is that genetic variation is random in Darwin, and directed in LaMarck. This difference between genetic variation is the difference between sustaining technology evolution and radical technology evolution.

Darwin identified a pre-Darwinian evolutionary period characterized by sexual reproduction, among unrelated species. In the pre-Darwinian period, the basic biochemical machinery of life evolves very rapidly.

Darwin's evolutionary model is based upon random variation, which results in two-parent cross breeding between two different product species. Darwin's theory was that offspring differ from their parents in ways that are purely random.

In Darwin, some of those offspring are better able to survive and to reproduce than others.

In Lamarckian inheritance, the genetic variation is directed by the genetics of the parent phenotype. The technology evolution is asexual product evolution. Lamarck said that evolutionary change was slowly introduced into the species and passed down through generations.

As two-parent cross breeding slows down, the environment begins to look more and more like LaMarck's asexual genetic variation. The subsequent evolutionary period is characterized by asexual, non-interbreeding species.

The LaMarckian period features very slow evolution because individual species, once established, evolve very little. Darwinian evolution requires species to become extinct so that new species can replace them.

Most of the time, the evolutionary mutation in sustaining product technology is very slow, much like a neutral genetic drift, as described by LaMarck.

In the case of asexual product evolution, there is intense competition for knowledge, as food, within the product species between older versions and the most recent technologically advanced version.

The intra-species competition, both for knowledge to eat, and for market selection to reproduce, hastens the time to product extinction.

In asexual, adaptive LaMarckian economic evolution, it is very difficult for an economy to break away from equilibrium because the genetic diversity declines with each passing generation of sustaining innovation.

Following Burger's analysis, "Natural selection acts in many different ways on the phenotype of the organism...In principle, selection can be described by a fitness function that relates fitness of individuals to the (quantitative) traits under selection." (The Mathematical Theory of Selection, Recombination and Mutation, 2000).

As the biological metaphor is applied here, products are phenotypes with certain technological genetic characteristics that are selected by consumers based upon the fitness function between the consumer's mental image of how the product will "fit" given the technological characteristics and traits of the product.

The selection in sustaining innovation follows Leven's biological model, where selection,

> operates separately within each local deme in a density-dependent manner: the fraction of adults in every deme is fixed...Nonrandom assortment of genotypes into demes, however, and certain spatial structures may easily lead to the maintenance of stable polymorphisms...(where) detrimental mutations may accumulate and eventually become fixed, thus leading to a progressive fitness decline that can result in population extinction."(Evolution in Changing Environments, 1968).

The greater the rate of selection of mass produced, standardized products in the global managed market, the greater the rate of product obsolescence and the greater the rate of economic decline to product extinction in the Nash Equilibrium.

Put another way, the more the offspring product, in an asexual genetic process looks like the parent product, the less likely genetic crossover will occur because of the greater likelihood that consumers will select the product that is slightly better than the old parent product, but not radically different.

As the global macro production technology becomes standardized, the multinational corporations abandon new radical technology and seek out advantages associated with tax incentives and skills that complement the corporate core technology.

The large corporations are engaged in political "rent-seeking" to replace the bigger rates of profit they obtained before the products entered the mass global market.

The political manipulation of government agencies is aimed at controlling the flow of knowledge, which if is horizontal, jeopardizes their flow of legacy technology profit.

The financial risk of the large corporations was accurately described by Christensen as getting "blown away'" by the introduction of a new radical product.

In any mass-produced product phenotype, the asexual product is in jeopardy of being displaced by its own offspring.

The greater risk is that the asexual product becomes obsolete (extinct) in a rich heterogeneous environment because of the greater likelihood of genetic crossover between product species.

Which explains the rent-seeking political behavior of large corporations in manipulation of government agencies and agents to hang on to their legacy income.

The large corporations seek to control the food supply of knowledge by constraining the flow of new information in a vertical, controlled flow of information.

John Maynard Smith describes one of the reasons why the flow of information and knowledge is critical to technology evolution.

> Information can pass from DNA to DNA, and from DNA to protein," said Smith. "But, information can not pass from protein to DNA – information being passed specifies the amino-acid sequence of proteins...if a protein with a new amino acid sequence is present in a cell, the protein can not cause the production of a DNA molecule with the corresponding base sequence. (Evolutionary Genetics, 1989).

Smith provides a technological metaphor to help readers better understand what he is describing. "You cannot make a new record by singing into the speakers of a stereo."

If the genetic information flow is one-way, then the biological equilibrium in one time period would not exactly be like the equilibrium of the ensuing time period.

In genetics, for example, the introduction of new information, in the form of a new molecule, would alter the protein structure, possibly resulting in a new genotype.

The new genotype, given the right environmental conditions, could possibly result, in an evolutionary way, a new radical phenotype.

Part of the complexity of understanding the effect of a one-way flow of new genetic information is that the effect depends on the environmental conditions at the moment the new information is introduced.

The effect on species in an asexual reproducing environment will be different than the effect in an environment where there is sexual reproduction, which may lead to the introduction of entirely new species.

The natural forces of transmission, selection, and genetic variation operate differently, both at the genotype and phenotype level in each environment.

In the case of an asexual reproducing environment, for example, the natural forces of evolution lead to stasis and decline because the forces of equilibrium overwhelm the forces that resist equilibrium.

In this case of asexual evolution, "breaking away from equilibrium" would not be expected.

Austin Hughes (1999), describes this asexual process as adaptive evolution.

> Thus, a process of "hitchhiking" occurs," says Hughes, "whereby a certain portion of the chromosome on which the mutation originated is carried along with it...linked nucleotide sites are "swept" to fixation along with a linked favorable mutual called "selective sweep.

"At the population level," adds Hughes, "this process will have the effect of reducing variation at sites linked to the locus under selection."

Carl Woese describes this idea as vertical gene transfer, and contrasts the asexual process with Darwin's "horizontal" gene transfer.

In horizontal gene transfer, according to Woese, the sharing of genes between unrelated species is prevalent. The horizontal gene sharing allows for new genetic information to be shared so that "clever chemical tricks and catalytic processes invented by one creature could be inherited by all creatures." (The Genetic Code: the Molecular Basis for Genetic Expression, 1967).

As applied to the technology metaphor, the heredity of product phenotypes from the parent's genes are transmitted in an asexual, single-parent, process of heredity.

As is the case in biological evolution, the deleterious mutations accumulate and usually end up as lethal to the organism, commonly called "inbreeding."

As a result of the asexual process associated with deleterious mutations, the phenotype's offspring do not obtain "good" genes.

Consumers, who are selecting the product phenotypes, base their decisions upon the "fitness" of the product with their own sovereign welfare function.

The deleterious mutations in the product accumulate, or become obsolete, and once they are obsolete, they never come back.

This outcome in asexual reproduction is the same conclusion reached by Woodruff and Thompson, who state,

> ...asexual reproduction is an evolutionary dead-end because it will lead to deterministic, open-ended mutation accumulation and eventual extinction. (Mutation and Evolution, 1998)

Consumer demand in the existing market is acting as the selection mechanism, and in the case of product evolution, this would be the case of adaptive, selection-induced mutation.

It is positive and directed selection, with genetic technological mutations accumulating in the technological characteristics of products that are being selected by consumers in the existing market.

Those products not being selected in the market, in other words, products who are not close to the "best-fit" technology, are deleted from the mutation process.

The fate of the original asexual product mutation is altered and affected by the appearance of a superior mutation, which ultimately replaces the original parent product, technologically, only to be replaced by some future product innovation, until the entire repertoire of technological genetic diversity within the economy is used up.

As described by Woodruff and Thompson, asexual product genetic mutation is an evolutionary dead end.

Chapter 5.
Disruptive Innovation As Two-Parent Product Genetic Crossover.

It is the very rare occurrence of "sexual" mating between two different product technologies that causes disruptive radical technology evolution.

The two-parent crossbreeding generally occurs along the technological "boundaries," between the two products that share some technology in common.

The crossbreeding between products "stirs up" the technology possibilities frontier by introducing new technological alleles into the economy. The new genetics pushes the technological possibilities frontier outward.

In contrast to the asexual genetic inbreeding process, sexual reproduction between two product parents allows "good" genes from both parents to be mixed, via a process called mutation crossover.

The mixing of good genes causes a slight selective market advantage for the new product because the prior product markets are composed of consumers in both product markets, who had seen something that looks like the new product before.

In order for two products to crossbreed, the technological chromosomes must be located close to each other on DNA strands, as would be expected in any industrial cluster. The radical product introduction into the market represents an evolutionary "genetic bifurcation."

As a consequence of technology crossbreeding, new future markets for the new product may be created. If a sufficient level of market demand materializes, then the existing market for the two parent products may bifurcate into an entirely new future market.

Bifurcation of product technology and the subsequent micro market bifurcation, is analogous to the appearance of a new species in Darwin.

When these new, or novel, environmental conditions occur, existing organisms in the environment are confronted with choices about how to adapt and survive.

The new product creates a novel event for humans to adapt to, and like all novel events, the human brain sorts and selects based upon prior memories and new mental images, trying to obtain quantum coherence in how to respond to the novel event.

The decisions by consumers and producers are not random, but purposeful, and consequently would not be modeled based upon Darwin's theory of random selection.

The consumer's response to the appearance of the novel event determines if *ex nihilo* preferences are created in the as-yet-to-be created future market, which may contingently be created, if a sufficient level of market demand for the higher priced novel product materializes.

Only in the case of sexual reproduction, when technology from two different technological frontiers combines genetic material, in a type of genetic technological crossover, are the conditions for new evolutionary trajectories created.

In Darwin, if one phenotype has an accumulation of deleterious mutations, and mates with a phenotype that does not have copies of the bad genes, then the offspring have advantageous mutations without obtaining the deleterious mutations.

Thus, in Darwin, a bad set of genes can be overcome, in an evolutionary process that allows the lethal trajectory to be reversed. The new species adapts and survives in the environment.

Sometimes, contingently, that radical product bifurcation leads to a market bifurcation, where the future final demand in the finished goods market place is completely different than the market demand in the previous time period.

The final demand markets bifurcate when the structural relationships between firms in the intermediate demand markets change to accommodate production requirements for the radically new product.

The new intermediate market relations disrupt and displace the prior interindustry relationship, which has a cumulative causation effect on future income distribution.

In other words, the product technology bifurcates first, which may lead to a micro product market bifurcation.

Then, much later, the entire macro economic market may bifurcate into a new market structure, with an entirely new macro economic demand structure.

The new macro market bifurcation causes prior income distributions to change. After income distributions change, the economy will never return to the prior equilibrium, because that former market does not exist, anymore.

Costs of production change, and consequently, prices on existing older goods decline at a faster rate, and the profit calculation of venture capitalists change.

Other species, in this case, humans, have to figure out what the appearance of the new product means for their survival.

In the existing final demand markets for finished goods, under the prior market relationship, there are two types of information searching activities being conducted by consumers and producers, as they attempt to figure out how to respond to the novel conditions.

First, the producers are searching for knowledge, which serves as food, or energy, for firms.

Some food gets eaten, some food gets stored for later use, and some food may spill over, accidentally, onto the plates of nearby firms, which had never eaten that kind of food before.

Second, the consumers are searching and sorting mental images, trying to select the right "fit" between the new product and their own individual welfare function.

The reason that technology evolution is positive, haploid and directional is that political and corporate elites want to direct who or what firm gets to eat the food, in a vertical or constrained way.

If the elites are successful in limiting the flow of knowledge to firms, they may be able to restrict the range of products available to consumers because knowledge is what spurs radical product innovation.

The elites attempt to control the flow of information to hang on to their legacy incomes and profits from the existing status quo market.

In order for genetic technological diversity to occur, the one-way flow of knowledge from idea to market, must be horizontal and free, not vertical and constrained by forces that inhibit the free flow of technological information.

When price and technological information is not horizontal and free, the economy cannot break away from equilibrium. In other words, the market cannot bifurcate, and new future income flows cannot be created.

The theory of technology evolution predicts that the greater the span of genetic technology in two-parent product mutation, the greater the risk associated with what consumers will do when they see the new product in the market for the first time.

Greater consumer risk, all other things being equal, requires a greater level of consumer trust, and greater allegiance to the rule of law, so that the consumer can have enough confidence to try out the new product.

The great uncertainty about what consumers will do when they first see the product creates great risk for producers in production of the new product.

Just like the level of trust and allegiance to the rule of law for consumers, the producers require a stable institutional arrangement in obedience to the rule of law, so that risks taken in period one can lead to appropriation of income in period two.

The consumer risk and the producer risk compound the investment risk taken by capitalists to make the initial capital investments in the new product.

The investment risk may not be reaped for 8 to 10 years, and capitalists must have a great deal of confidence in the stability of the system of justice, over an 8 year period of time, in order to appropriate their capital gains.

The theory of technology evolution of two-parent crossover is described in Diagram 10.

The Diagram describes two distinct types of crossover between two parents.

The first stage of genetic crossover is the technology crossover.

The two products are located close to each other, and share some technology, both in the production processes and in the product genetics.

The professional scientific staff, engineering staff, sales staff and plant floor supervisors meet each other in professional meetings or other social business networks.

Mostly, they talk about how to solve the problems they are having in production and marketing.

From a theoretical perspective, what they are doing is sharing and diffusing technological knowledge.

The staff which returns from the meetings are called "boundary-crossers," because they cross industry and product boundaries to bring back technological knowledge to their firm.

In most cases, the boundary-crossers contribute to sustaining innovation in existing products by making small adjustments in production and technological features for their existing firm.

In very rare cases, the boundary-crossers may collaborate with staff from other firms on the creation of entirely "new-to-the-world" products.

The second type of crossover described in Diagram 10 occurs much later in time as "market crossover," when consumers from both of the parent products take a risk in buying the new product.

One important part of the theory of technology evolution concerns the vital, primary role of the consumer selection process.

Technology evolution is primarily a market driven, not a technology driven, evolutionary event.

Diagram 10 shows that yellow-green parents created a blue offspring product.

Diagram 10. Depiction of Two-Parent Product Genetic Crossover.

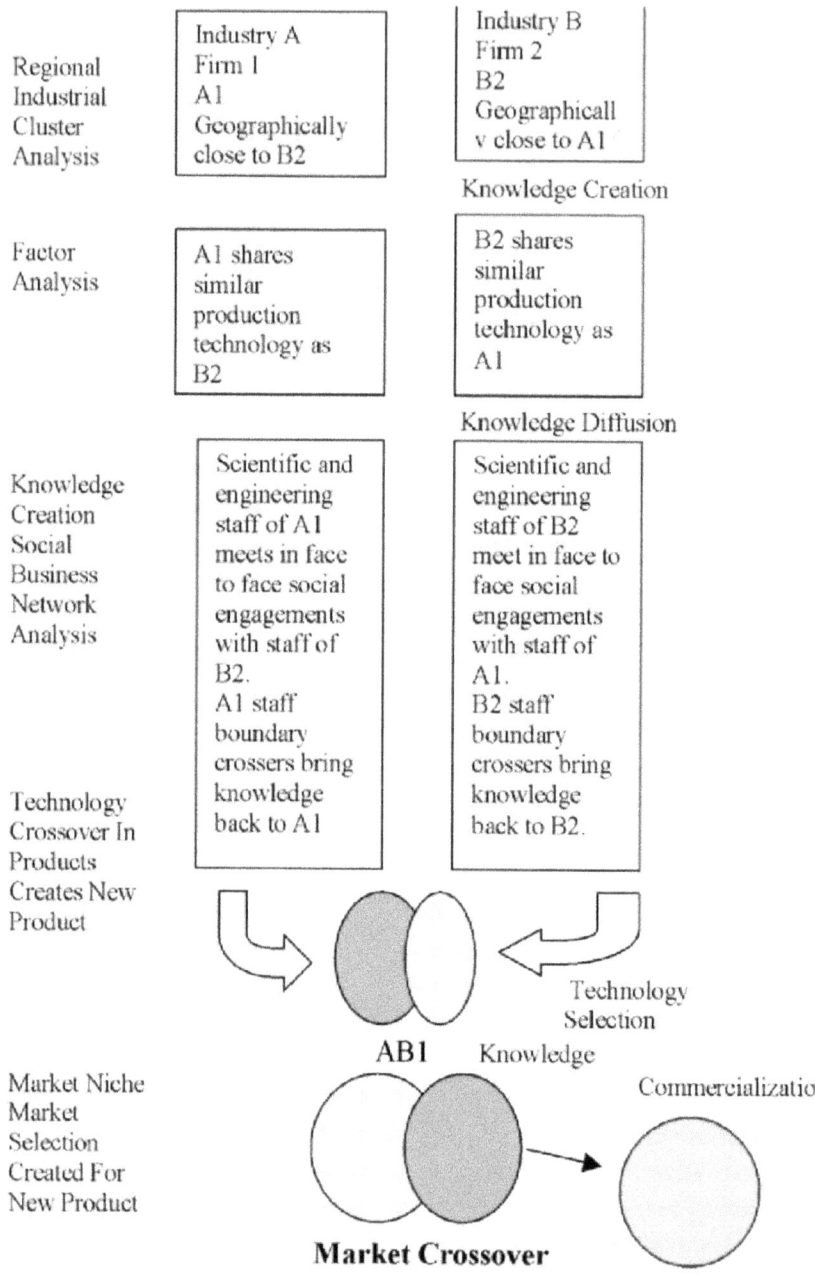

Regional Industrial Cluster Analysis	Industry A Firm 1 A1 Geographically close to B2	Industry B Firm 2 B2 Geographicall v close to A1

Knowledge Creation

Factor Analysis	A1 shares similar production technology as B2	B2 shares similar production technology as A1

Knowledge Diffusion

Knowledge Creation Social Business Network Analysis	Scientific and engineering staff of A1 meets in face to face social engagements with staff of B2. A1 staff boundary crossers bring knowledge back to A1	Scientific and engineering staff of B2 meet in face to face social engagements with staff of A1. B2 staff boundary crossers bring knowledge back to B2.

Technology Crossover In Products Creates New Product

Technology Selection

AB1   Knowledge

Market Niche Market Selection Created For New Product

Commercializatio

Market Crossover

If these two types of crossovers occur, the theory of technology evolution predicts that if,

- A new two-parent product is imagined and visualized, and if,
- A new product prototype is created, and if,
- The new potential product can find capital to produce and market the product, and if,
- The new radical product can find a small consumer niche to buy the higher priced product, when the consumers first see it,
- Then, there is a slight possibility that the new product phenotype can survive, without being killed by the incumbent firms.

If the new product is commercialized, it may create new inter-industry intermediate demand relationships between the suppliers from both the parents.

The new product may share some parts of the consumer demand market that would contribute to the creation of a small niche market.

Diagram 10 is adjusted as Diagram 11, to describe the evolution of the smartphone, which crossed technology from a cell phone with a computer chip, and a digital camera.

Both the cell phone and the digital camera had undergone a series of sustaining innovations, prior to crossbreeding, in 2000, to create the smartphone.

The parent product of the cell phone was a two-way mobile radio, created in 1941, by Galvin Manufacturing Corporation, later to be named Motorola Corporation..

Credit. Motorola Corporation.

The radio was successfully commercialized when the City of Philadelphia took a risk in purchasing the radio for police cars.

The two-way radio used a spectrum of the radio frequency that allowed the radio to communicate, without radio interference, in private.

About two decades later, the inventor of the first version of the cell phone used electronic parts from the Motorola two-way radio to create a hand-held cell phone.

About a decade later, on April 3, 1973, Motorola publicly demonstrated the world's first portable cellular telephone and system, called the Motorola DynaTAC (DYNamic Adaptive Total Area Coverage).

The Motorola DynaTAC phone, received approval to sell the new phones from the U.S. Federal Communications Commission on September 21, 1983. The 28-ounce phone became available to consumers in 1984.

Motorola engineers had been experimenting with the internal electronics of the mobile radio since the first hand-held cell phone in 1960s. The experimentation is consistent with the theory of technology evolution as a sustaining innovation.

This version of the cell phone was mated with the technology of a digital camera, which had also been undergoing its own sustaining evolutionary trajectory.

In 1972,Texas Instruments, patented a film-less electronic camera, which became the second parent product of the smart phone.

The Texas Instrument product underwent a series of sustaining innovations, which made the camera more user-friendly.

In 1975, Steven Sasson , an engineer with Eastman Kodak, combined electronic parts from a Motorola mobile radio with a Kodak movie-camera lens and some newly invented Fairchild CCD electronic sensors to make an early version of the digital camera with greater functionality.

In August, 1981, Sony released the Sony Mavica, the first commercial electronic camera. Images were recorded onto a mini disc and then put into a video reader that was connected to a television monitor or color printer.

Credit. Sony Corporation.

In 1990, Kodak developed the Photo CD system and proposed "the first worldwide standard for defining color in the digital environment of computers and computer peripherals."

The technology standardization in computer peripherals paved the way for the two-parent technology crossover between the cell phone and the digital camera.

Before that technology crossover occurred, IBM created a multifunctional hand held personal digital assistant called Simon. The IBM Simon Personal Communicator, released in 1994, sold for a very high price of $1,100.

In addition to its ability to make and receive cellular phone calls, Simon was also able to send and receive faxes, e-mails and cellular pages.

The Simon quickly became extinct, having sold a total of 50,000 units.

The consumer's response to the appearance of Simon describes what happens when a radical new product does not create *ex nihilo* preferences.

The new product must develop a small niche market, that grows, ultimately, into a mass, global market.

In 1995 Kenneth Parulski and James Schueckler, two engineers at Kodak, patented a camera phone.

The patent application specifically described the combination as either a separate digital camera connected to a cell phone or as an integrated system with both sub-systems combined together in a single unit.

The technology crossover to the smartphone occurred in 2000, when Ericsson released the R380.

The R380 was the first product to be officially billed and marketed as a smartphone.

Diagram 11. The Technology Evolution of the Smartphone, That Crossed Technology From a Cell Phone With a Computer, Access to the Internet and a Digital Camera.

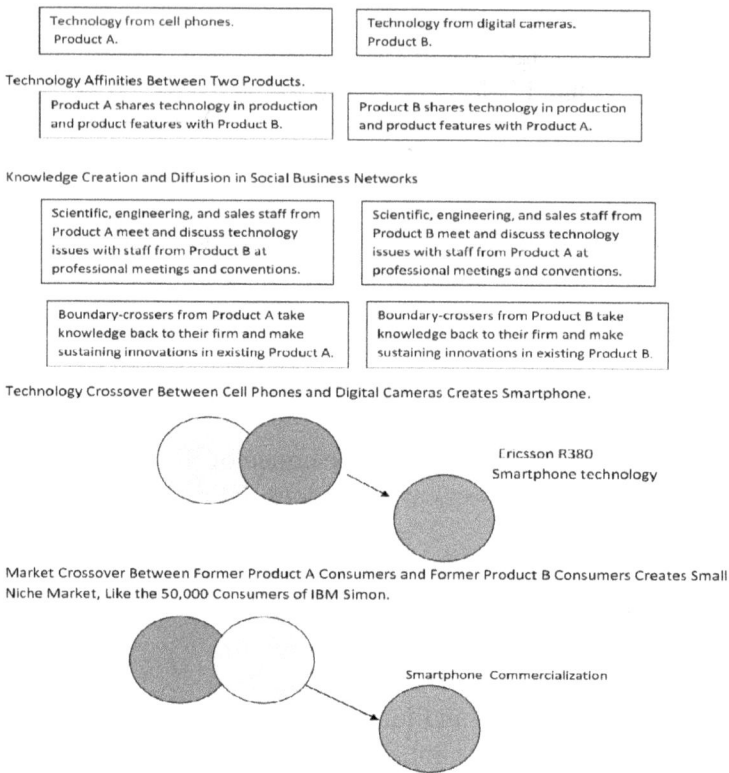

| Technology from cell phones. Product A. | Technology from digital cameras. Product B. |

Technology Affinities Between Two Products.

| Product A shares technology in production and product features with Product B. | Product B shares technology in production and product features with Product A. |

Knowledge Creation and Diffusion in Social Business Networks

| Scientific, engineering, and sales staff from Product A meet and discuss technology issues with staff from Product B at professional meetings and conventions. | Scientific, engineering, and sales staff from Product B meet and discuss technology issues with staff from Product A at professional meetings and conventions. |

| Boundary-crossers from Product A take knowledge back to their firm and make sustaining innovations in existing Product A. | Boundary-crossers from Product B take knowledge back to their firm and make sustaining innovations in existing Product B. |

Technology Crossover Between Cell Phones and Digital Cameras Creates Smartphone.

Ericsson R380
Smartphone technology

Market Crossover Between Former Product A Consumers and Former Product B Consumers Creates Small Niche Market, Like the 50,000 Consumers of IBM Simon.

Smartphone Commercialization

The radical new product combined the technology of the cell phone, digital camera, computer personal assistant, and access to the internet.

Initially, the niche market for the R380 developed slowly.

However, in Japan, consumers had seen something like the R380 before, and were willing to take a risk on buying the new higher priced product.

By the end of 2001, in Japan, there were more than 40 million subscribers of NTT DoCoMo service nationwide.

At the end of 2015, over 400 billion smartphones had been sold, globally.

The theory of technology evolution predicts that when the new radical product enters the mass, global market, that sales revenues will decline.

It took the smartphone about 15 years to use up the genetic material from the two parent products.

In the mass, global market, unit sales may be stable or declining, but the marginal profit per unit is declining.

At the point of declining marginal profits, the product has used up the genetic diversity in the technological possibilities frontier and will not benefit from future sustaining innovation.

The smart phone is following the evolutionary path to technological obsolescence and product extinction, as depicted in the declining global unit sales, in Diagram 12.

According to sales statistics presented by IDC:

> Smartphone vendors shipped a total of 355.6 million units worldwide during the third quarter of 2018 (Q3 2018), resulting in a 5.9% decline when compared to the 377.8 million units shipped in the third quarter of 2017. The drop marks the fourth consecutive quarter of year-over-year declines for the global smartphone market.

Diagram 12. Global Unit Sales of Smartphones by Top 5 Vendors.

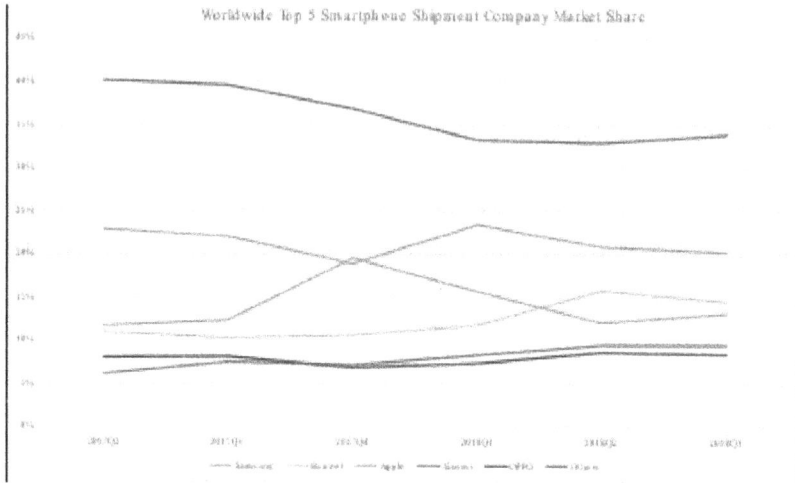

Worldwide Top 5 Smartphone Shipment Company Market Share

Credit IDC

In the U. S., employment in the smartphone industry has been declining since 2008.

Diagram 13. U. S. Total Employment in the Smartphone Industry, 2008 to 2018.

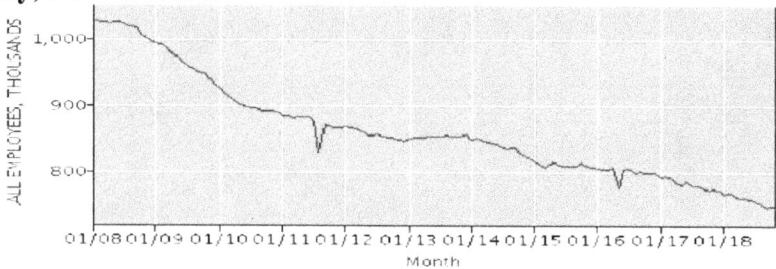

Super Sector: Information
Industry: Telecommunications
NAICS Code: 517
Employment, Hours, and Earnings from the Current Employment Statistics survey (National)

Diagram 14 explains the technology evolution of the smartphone from the theoretical perspective of the biological metaphor.

In the 1980's to 2000, technological genetic diversity was increasing, leading the economy to a new attractor point that featured greater incomes and employment.

On the upward trajectory of growth, the smartphone created huge new flows of profits and income, leading to the future attractor point.

After the smartphone entered the mass, global market, around 2008, most of the genetic diversity gained in the two-parent genetic crossover had been used up.

From the new attractor point, shown in Diagram 14, the smartphone must find something new to stir up technological diversity, or the market demand for the product will slide down through Muller's Ratchet, and stabilize at a Nash Equilibrium, as depicted in Diagram 15.

Diagram 14. Application of the Biological Metaphor To The Example of the Two-Parent Crossbreeding in the Smartphone.

GENETIC TECHNOLOGICAL CROSS BREEDING IN TWO PARENT PRODUCT ENVIRONMENT

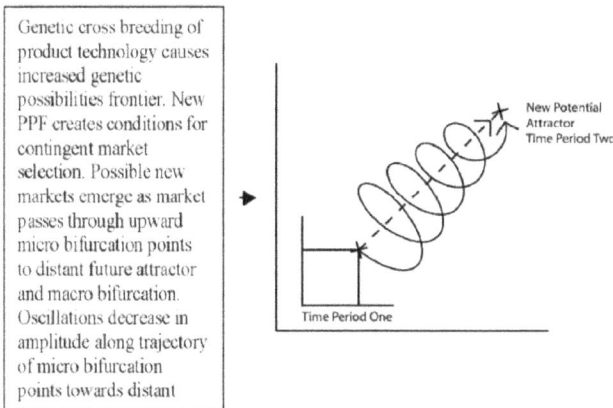

Genetic cross breeding of product technology causes increased genetic possibilities frontier. New PPF creates conditions for contingent market selection. Possible new markets emerge as market passes through upward micro bifurcation points to distant future attractor and macro bifurcation. Oscillations decrease in amplitude along trajectory of micro bifurcation points towards distant

New Potential Attractor Time Period Two

Time Period One

Diagram 15. Depiction of Smartphone Future Trajectory Without Future Genetic Diversity.

ASEXUAL SINGLE PARENT ENVIRONMENT

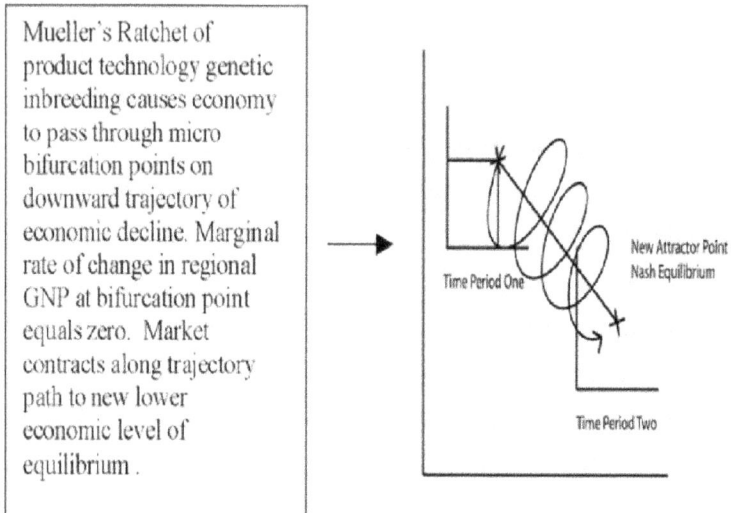

Mueller's Ratchet of product technology genetic inbreeding causes economy to pass through micro bifurcation points on downward trajectory of economic decline. Marginal rate of change in regional GNP at bifurcation point equals zero. Market contracts along trajectory path to new lower economic level of equilibrium .

Time Period One

New Attractor Point
Nash Equilibrium

Time Period Two

In the application of the biological metaphor to technological evolution, the consequence of the existing level of technological product genetic diversity in the smartphone can lead to three possible future states.

As described in Diagram 16, in the prior period of time, before the two-parent crossover occurred, is characterized as the antecedent economic and financial environmental conditions.

Diagram 16. Adaptation of Biological Metaphor to Three Future States of the Economy.

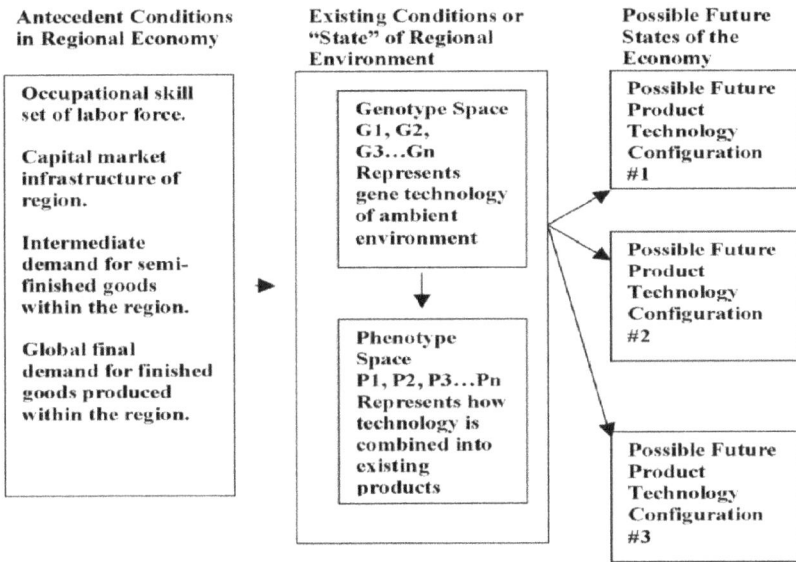

Antecedent Conditions in Regional Economy	Existing Conditions or "State" of Regional Environment	Possible Future States of the Economy
Occupational skill set of labor force. Capital market infrastructure of region. Intermediate demand for semi-finished goods within the region. Global final demand for finished goods produced within the region.	Genotype Space G1, G2, G3...Gn Represents gene technology of ambient environment ↓ Phenotype Space P1, P2, P3...Pn Represents how technology is combined into existing products	Possible Future Product Technology Configuration #1 Possible Future Product Technology Configuration #2 Possible Future Product Technology Configuration #3

The prior technological conditions evolved into the current, or existing, economic conditions, around 2008, characterized by the phenotype and genotypes of the product technology.

From the existing time period, the smartphone can trace out an upward trajectory to a future attractor point, based upon a second two-parent genetic crossover.

Or, the smartphone can trace out a downward trajectory to a Nash Equilibrium, in an asexual genetic sustaining innovation environment.

Or, the most likely outcome for the smartphone is genetic status quo, as a result of the political and financial manipulation of information and knowledge flows by corporate elites and government agencies.

Chapter 6.
The Relationship Between Capital Investments and Technology Evolution.

The key causal variable in technology evolution, prior to a two-parent genetic crossover, is an investment of capital directed to the new radical product.

The sources of capital face great uncertainty and risk on making an investment in a radical new product.

The three unknowns that are creating the investment risk are:
- What will consumers do when they first see the new product?
- Can producers combine the production technology in the two products in order to mass produce the product?
- Will the incumbent products kill the new species before it can be commercialized?

The capitalists, in time period one, must make several sequential guesses about the future, and if they guess right, there is a slight possibility that the capitalist can make a profit.

They must first guess how the technological properties of a new product are linked to the consumer preferences, via product heuristics in the final demand market for finished goods.

Second, they must guess what prices the new product may sell at in the future, in order for them to guess at their potential future rate of profit in time period two.

Then, they must guess whether that future rate of profit meets their own internal preferences of capital gains, given all the other possible capital investments that they could make today.

The key to understanding the trajectory of the economy is that the capital investment in time period one must be directed to creating genetic diversity, which means capital investment in radical new two-parent products.

In the absence of an investment in two-parent crossover, the economy is on a trajectory to a Nash Equilibrium. This insight into the theory of technology evolution is similar to the Keynesian insight into a macro economic equilibrium at a low level of aggregate demand.

As Keynes noted in his 1930 work, Treatise on Money, an economy that does not have an adequate savings/investment relationship to prices and profits, leads to a future level of aggregate demand that does not fully employ the resources of the economy.

Keynes explained that a national economy could be stuck at the lower level of aggregate demand in time period two, if the savings/investment relationship was not adequate in the prior time period.

In Keynes, part of the supply of investment capital is related to profits from existing firms, in time period one. The source of capital for Ericsson, for example, in developing the smartphone was from internal cash flow.

The competitors to Ericsson entered the market by investing capital derived from internal operating revenues, not from outside sources of capital.

Those profits made by Ericsson acted like a beacon of light in attracting other potential competitors to enter that market.

If the potential competitors "see," or discover that profits are being made, they will enter the market, using their own internal sources of capital, derived from cash flow.

The competitors who entered the smartphone market had to guess if making a new investment in smartphones was more profitable than making a sustaining capital investment in their own existing products.

This decision by the 4 competitors is an example of Christensen's innovator's dilemma.

The decisions of all the potential competitors was constrained by their level of corporate debt, in the time period around 2000.

Commercial bankers and corporate bond holders act to limit the additional capital investment of corporations through restrictive covenants on what a corporation may do with its free cash flow.

Wassily Leontief, in Essays In Economics, Theories, Facts and Policies, (1977), expanded Keynes' sources of capital by describing that source of capital for investing in genetically new products was a combination of "entrepreneurial returns, monopolistic revenues and windfall profits resulting from appreciation of commodity stocks on hand."

In his theory of economic equilibrium, Leontief described how the economy was a circular flow of capital

Leontief pointed out that the entire combination of capital, called "undistributed surplus," plays a double role for the firm.

"On the one hand," Leontief said, "it represents the values of materials, additional pay rolls, and other cost elements, on which, it usually is spent in the process of new investment."

The entrepreneurial profits, represent "on the other hand, the productive services of capital and entrepreneurship."

The capital gain on capital invested, according to Leontief, depended on conditions situated outside the sphere of production. The economy represented a circular flow of capital when this entrepreneurial gain was reinvested back into the economy.

In contrast to Keynes and Leontief, in the theory of technology evolution, the source of capital for making investments in a radical new product are profits from a venture capital exit event, in time period one.

The units of capital associated with venture capital exit events are not related to Keynes' analysis of production prices in the existing market, for existing goods.

Nor are the capital sources based upon the interest rates applied to the repayment of debt from loans made by commercial bankers.

Joseph Schumpeter, in The Theory of Economic Development, (1961), explained that,

> capital gain profits, from earlier ventures, are the source of funds for later ventures, and those profits are not, in any way, derived from the price mechanisms in the sphere of production described by Keynes.

For Schumpeter, as in Ricardo, income distribution in the earlier time period explains, or determines, prices in the later time period for both capital equipment and the products that the capital equipment produces.

Schumpeter explained that one effect of capital investments in new ventures is to cause new distributions of wealth among society.

"It is not essential," says Schumpeter,

> that the new combinations should be carried out by the same people who control the production or commercial process...new combinations mean the competitive elimination of the old...It (the investment process in new firms) explains on the one hand the process by which individuals and families rise and fall economically and socially...In a non-exchange economy, for example, a socialist one...the social consequences (of the investment process) would be wholly absent.

The capital investment in new firms creates new income flows in the future, which Schumpeter called "purchasing power," at the same time that the new ventures eliminate the existing firms.

Schumpeter stated, "The entrepreneur needs capital...to serve as a fund out of which productive goods can be paid for...it is a fund of (future) purchasing power."

The venture capitalist, who obtained the profits from the earlier exit events, is continually looking for new firms in time period two to invest in.

The entrepreneurs who create new ventures, for Schumpeter, represent the demand for capital, while the venture capitalist represents the supply of capital.

Schumpeter wrote,

> The (new) demand...causes a rise in the prices of productive services. From this (rise in prices) ensues the withdrawal of good from their previous use...the newly created purchasing power (new incomes) is squeezed out at the cost of previously existing purchasing power...

The main point made by Schumpeter is that capital gain profits from earlier venture capital investments are the source of funds for later ventures, and in this process, the future environment is replicating itself by reproducing new ventures.

In Schumpeter, the evolutionary effect of capital investment arises because,

> new enterprises do not grow out of the old, but appear alongside of it, and eliminates it (the old enterprise) competitively, so as to change all the conditions that a special process of adaptation becomes necessary.

These new firms enter the economic fabric at a specific moment in time, and according to Schumpeter, the economic fabric at that moment is seen as a "... circular flow of economic life."

Schumpeter describes a type of market bifurcation that results from the capital gain investments by stating that,

> ...a discontinuous change in the channels of the (circular) flow... disturbance of equilibrium, which forever alters and displaces the equilibrium state previously existing...These spontaneous and discontinuous changes in the channel of the

circular flow and these disturbances of the center of equilibrium appear in the sphere of industrial and commercial life, not in the sphere of the wants of the consumers of the final product.

Schumpeter explained that capital investments in radical new products is opposed politically, by powerful economic and political interests who benefit from the initial distribution of income.

Those financial special interests have an allegiance to the status quo of production and income.

The relationship between politics and cultural values and the level of capital investments in radical new ventures is one reason why Darwin's theory of random variation and natural selection is different than the evolution of technology.

The selection of products is influenced by cultural and social values.

The selection process of politics and cultural values in an economy affects the supply of capital available for investments targeted to radical new products.

Under one set of cultural values, characterized by free, horizontal flows of information, capital can replicate itself through the addition to capital from exit events in prior new ventures.

Under another set of cultural values, the powerful forces of the status quo arrangement of income can redirect capital investments to a constrained and politically manipulated rate of capital investments, primarily in a limited range of sustaining, or asexual investments, aimed at improving existing production.

In his historical review of the capital that fueled the industrial revolution, Robert McAdams noted that historians share a widely-held belief about the origins of venture capital in the eighteenth century. (Paths of Fire: An Anthropologists Inquiry Into Western Technology, 1996).

McAdams wrote,

> ...all of the sources are in agreement that considerable increases in disposable wealth came into the hands of a substantial elite during the later part of the eighteenth century. Perhaps it is not so much their diversified and growing desires of consumers that quickened the pace of technological advance (during that era) but rather the increasing supply of potential venture capital for which this elite was beginning more aggressively to seek new avenues for profitable investment.

McAdams highlights two important factors about the origin and disposition of venture capital. First, he describes venture capital in the hands of the financial and political elite as "disposable wealth," not disposable income.

In the case of venture capitalists, the general source of wealth is the capital gain they achieve on the sale of their ventures.

Second, McAdams noted that the elites in the eighteenth century were looking for new ventures to invest their wealth. The elites were venture capitalists, not commercial bankers.

The pool of capital that fueled the industrial revolution did not depend on cash flow from old production units, nor is it related to the conventional loans made by commercial bankers.

The potential pool of capital, in the industrial revolution, was created by profits on exit events made by the venture capitalists, just as it is today for capital investments in radical innovation.

The new ventures that the venture capitalists create act as a market signaling device, in an information feedback system in the intermediate demand markets that function to coordinate mutually reinforcing expectations between the investors and the consumers.

In the theory of technology evolution, the explanatory variables of capital investment in time period one are:

- Rate of venture capital profits from "exit" events.
- Rate of venture capital and angel investments in new products.
- Rate of institutional capital investment in new production technology.
- Birth rate of new ventures producing technologically new products.
- Birth rate of new ventures producing newer versions of older products.
- Rate of old product obsolescence and life-cycle.
- Death rate of companies producing older products.
- Death rate of new ventures creating new technological products.
- Rate of mass layoffs in multinational branch plants located in an economy.
- Rate of new jobs created in multinational branch plants located in a region.
- Rate of commercial or other institutional bank loans to existing companies.
- The element of trust and joint expectations, in other words, joint and common beliefs about the future, in a society characterized by the rule of law.

Economies and societies that are based upon the rule of law, coincidentally, create an environment most conducive for the creation and diffusion of knowledge.

Chapter 7.
Knowledge Creation and Knowledge Diffusion In Technology Evolution.

New knowledge is created in a mental "insight-imagination" process when the brain encounters a novel circumstance.

This mental process is commonly called "imagination."

The future that is imagined is tried out mentally, thousands of times a second in an individual's brain as it sorts and searches for what may happen if a sequence of events occurs.

As the image of the novel condition is filtered and processed by hundreds and thousands of neural synapse firings, the brain is searching and sorting for the right fit between the external image and the internal model of the image.

The main goal of the sorting and searching, from a biological perspective, is quantum coherence in a course of action or behavior to seek advantage or seek survival.

The brain is calling upon memory circuits of something that looked liked the novel circumstance, and combining those memories with new internal images to come up with the right fit and the right course of action.

The new knowledge takes the form of new mental constructs and neural connections in the brain.

New knowledge is created as a result of the confirmed beliefs associated with an individual's internal and external images lining up together.

The knowledge creates a permanent construct in the brain, so that the individual will "know" what to do if the novel event occurs again.

As the thousands of images are sorted, the neural networks tend to line up and fire in sequence, which is a condition called "making up your mind."

Humans are confronted with novel circumstances primarily in their dealings with other humans.

Evolution of the brain has allowed humans to anticipate the response of other humans, and imagine what the other humans may do in the future, as if it had in fact occurred.

Applied to economics, humans in a market are continually imagining what other humans are going to do, and hypothesizing about the future. Human brains are always spinning out new alternatives and new ideas about the potential behavior of other humans in the market.

When the brains of other humans "see" the same novel event, and change their own images and insights, then the reality of the new knowledge is confirmed as "real," by the humans who are imagining the same future.

Scientists call this confirmation of realty between different humans as the "intersubjective" verification of reality.

Humans, in a technological community, interact with each other every day, by exchanging technical information on production and markets, based upon their store of knowledge and memories.

New knowledge in technological evolution seems analogous to food, or energy, which gets "eaten" in the Lotka-Volterra biological setting.

In the biological metaphor, the boundary-spanners are seeking food in the form of market and technical knowledge in their environment.

Translating to the biological metaphor, if the boundary-spanner gets back to her firm, and if the firm "imitates" the best production practices of other firms, the firm is "adapting" to the new knowledge in the environment.

In other words, what happens when the boundary-spanner gets back with knowledge is that the senior executives at the firm may begin "imagining" a new reality of how a new product, or new production technique, may fit into the firm's future market.

If the senior executives in the firm, who are knowledge-bearing agents, actually change their mental images to accommodate the new knowledge, there is a small possibility that they may decide to create a new product.

Creating knowledge and diffusing knowledge is not an automatic, or given function. The creation and diffusion of knowledge is a deliberate and intentional behavior.

The creation of new technical knowledge is a social phenomenon caused by the interaction of humans engaged in social/business relationships.

Knowledge creation is a social, cultural process and takes place in the institutional social networks in the economy.

The individuals in the firms communicate with individuals in other firms, at continuing education and professional trade meetings.

Within this social network, the boundary crossers engage in the biological activity of searching and selecting the food of knowledge for gaining advantage or for survival.

Knowledge, in its economic context, is created, and is diffused among a distinct economic population.

In this depiction of knowledge creation, the scientists, engineers, sales staff and service repair technicians are the agents who "bear" knowledge, and cross boundaries.

These agents consume knowledge, and their pursuit of knowledge becomes an important activity in understanding the evolution of technology.

Rinaldo Evangelista, explains, in Knowledge and Investment: The Sources of Innovation In Industry, (1999),

> The general process of technological change can be conceptualized as a process of generation of new technological knowledge as distinct from the process which leads to its actual use in production...in the form of new or improved machines, technical devices and operating systems.

New products, created by new knowledge, represent, in the Darwinian context, a new species, that are mutations of old genetic technology.

To the extent that a firm can imagine, and produce, a new product, there is a slight possibility that new knowledge will be created.

The units of analysis in the measurement of knowledge are the knowledge-bearing humans, who carry knowledge around in their heads.

Genetic technological diversity comes from knowledge creation and diffusion, which is then contingently commercialized, in two-parent genetic crossover.

Knowledge, in the theory of technology evolution, builds upon what exists in the memories of humans, and knowledge is cumulative, in the sense that it builds upon existing memories.

The social networks that exist within the economy, serve as the institutional framework for knowledge diffusion.

Knowledge, unlike autonomous price-based transactions in neoclassical tradition, does not start over from scratch with each new encounter between boundary crossers.

New knowledge is diffused within the social networks when the boundary crossers meet each other, in face-to-face encounters. The reason that face-to-face engagements are so important to the theory of technology evolution relates to how humans interpret each other's behavior.

It is the very subtle change in voice inflection and tone, the slight change in the eyes, and facial expression that convey the new knowledge.

The diffusion of knowledge is not random, but Lamarkian, in the sense that the search for knowledge is purposeful and results in behavior adaptations that affects technological genetics.

"When people meet, they communicate," wrote Ilya Prigogine, and "when they leave the meeting, they keep the memory of their encounter. When they meet other people, this communication is propagated to an ever-increasing number of participants." (Exploring Complexity: An Introduction, 1989).

In Theoretical Welfare Economics, (1957), Jan de Van Graaff made this same point about the the social process of diffusing knowledge.

Graaff states that:

> ...the ultimate repositories of technical knowledge in any society are the men comprising it, and it is just this knowledge which is effectively summarized in the form of a transformation function...new knowledge, created with perhaps one purpose in mind, but is in fact valuable in a very different context.

Rikard Stankiewicz, in Technological Systems and Economic Performance: The Case of Factory Automation, (1995), notes that:

> Every engineer is embedded in a particular technological tradition characterizing his profession, the company he works for and the team he is a part of- the technological community...The accumulation and transmission of knowledge occurs in, and through, the formation of technological communities, and is strongly affected by their structure and dynamics.

The individual engineer "bears" knowledge, and works for a firm that is producing the range of product phenotypes within a region's "technological community."

What the engineer does with his knowledge, and how the knowledge is diffused within the community, become important elements of the theory of technology evolution.

Knowledge diffusion looks like the spread of a disease or infection throughout a population.

Epidemiologists call this diffusion of a disease a contagion, and speculate that the density of the humans in the human's habitat influences the rate of contagion.

There is both a feed forward effect from the production side, and a feed back effect in knowledge diffusion from the market demand side, and in this sense, knowledge diffusion exhibits a certain type of "Lamarkian" characteristic.

The diffusion of knowledge is not random, but Lamarkian, in the sense that the search for knowledge is purposeful and results in behavior adaptations of the "units of analysis" that affect genetics in the immediate time period following the adaptation.

Some economies have social networks that share norms and culturally-based systems of interpretation, where the deliberate behavior of knowledge creation occurs.

Those types of societies have a competitive advantage in the creation and diffusion of knowledge, compared to a region where the social networks are based upon other cultural values or norms of doing business.

If knowledge can be defined as an economic asset, or a pool of knowledge, then that pool or asset must be measurable, and additions to knowledge must be distinct.

Unlike natural resources, however, like a region's deposits of iron ore, the resource of knowledge can grow, and can also decline.

Knowledge, as a economic asset, is an essential precursor for technology evolution.

In this case, at the beginning of the time period one, an economy could be characterized as having a "stock" of knowledge based upon the number of agents in the economy who are engaged in searching for knowledge.

An initial stock of knowledge could either "flow" into a greater base of knowledge in time period two, or in the case of loss of genetic diversity, the stock of knowledge may "leak" out of the economy.

Sometimes, this phenomenon of knowledge leakage is called "brain drain," to refer to the idea that the agents who bear knowledge have left the region.

Applying the biological metaphor to economic evolution, the economy's "stock" of knowledge is contained in the structure of firms and population of technical staff in the region who are knowledge-bearing workers.

When those knowledge-bearing workers are free to communicate with each other, new knowledge may be created.

The genetic diversity of the knowledge grows when knowledge-bearing workers from one cluster of technology add to knowledge in another cluster of technology.

If new knowledge is not created, the economy is on a path to the Nash Equilibrium.

If the new knowledge were subsequently used to produce and deliver a new product to the market, the verification of the reality would be represented by a new production coefficient in the intermediate supply chain, signaling the entry of a new interindustry relationship for the new product.

The new interindustry relations replace the older social institutional structure of knowledge creation.

The act of creating the new product idea is captured in a moment in time in an economic accounting system by the appearance of a new coefficient in an input-output transaction matrix.

The appearance of a new coefficient in the matrix is, biologically, just like the appearance of a new neural synapse in the brain resulting from an insight/imagination that became new knowledge.

The new coefficient shows linkages between firms that allow for transfers and exchanges, that further confirm the new reality of the new knowledge.

The appearance of the new coefficient is a surrogate variable for new knowledge being created. In other words, the priced-based transaction coefficient is describing a non-priced knowledge transaction.

The linkages between firms are a result of something new occurring, and, that new thing is new knowledge based upon shared insights and imaginations of many humans.

In the sphere of knowledge related to production processes, the rate of adoption/imitation is affected by the closeness of the correlation coefficients in the transactions matrix.

Edward Feser's input output transaction table, with the newly added correlation coefficients in the intermediate sales matrix, allows for this complementary market development to be tracked and investigated. (Enterprises, External Economies and Economic Development, 1998).

The development of the intermediate market, as evidenced by new coefficients, is a surrogate model of how knowledge creation and diffusion is proceeding in the institutional networks of a specific economy.

The face-to-face exchanges in the social networks lead to tacit knowledge creation.

The characteristic of tacit knowledge provides the concept with a geographical boundedness of knowledge diffusion, because tacit knowledge in created through face-to-face communications.

In terms of the evolutionary metaphor, tacit knowledge is the knowledge carried around in the memories of the brains of humans. That tacit knowledge, in the biological metaphor, is food.

Biologically, tacit knowledge by the boundary-spanners in economics would look like bees pollinating flowers in nature.

The pursuit of knowledge creates a non-priced dynamic of competition, which involves how the agents deploy, or use their knowledge, in the marketplace.

As the agents diffuse the knowledge, it becomes known to others in the economy's institutional social framework.

Tacit knowledge that results in the creation of a radical new product creates an empirical marker as a coefficient in an input output transaction matrix.

In contrast to tacit knowledge, the other form of knowledge creation and diffusion is codified knowledge.

Codified knowledge is in the form of books, and manuals, or other documents, where the knowledge has been written down, or "codified."

Codified knowledge does not create new coefficients in the transaction matrix, it only modifies the size of existing coefficients.

The codified knowledge is easier to control than tacit knowledge because the elites can control who has access to the codified knowledge.

The exclusive reliance on codified knowledge, inside the firm, acts to limit the application of new technology, even if a boundary-spanner, or some other source of knowledge, happened to find its way into the firm.

Reliance on codified knowledge leads to economic "path dependence" because the application of codified knowledge constrains the choices of technology and the choices of people to imagine new opportunities.

An economy becomes "locked" into a pathway because the old ways of doing things become entrenched, and new knowledge, new ideas, new ways of doing things are limited by the reliance on codified knowledge.

In economics, this condition would be called constrained maximization within a production possibilities frontier that can never change.

From a biological evolutionary perspective, the reliance on codified knowledge in the firm is contributing to the creation of conditions of sclerosis and homogeneity in genetic production technology, as a result of this allegiance to the technological specialization.

The only outcome of codified knowledge is increased production efficiency, which leads to reduced costs, reduced profits, and ultimately, economic extinction.

The exclusive reliance on codified knowledge leads an economy to create a "specialization" in either production technology or products.

In a specialized, tightly bound economy, the most efficient "best practice" technology in the region evolves rapidly to the point of zero profits as more and more firms imitate the best manufacturing practice.

Specialization is merely genetic substitution in the existing environment's genetic technological possibilities envelope.

Specialization and sustaining innovation reshuffles existing DNA along the economy's production curve. The reshuffling and sustaining innovation produces greater production efficiency, but no genetic mutation.

Specialization results in a loss of genetic heterogeneity in technology, which means a loss in the economy's ability to create knowledge, which ultimately leads to an economy on a trajectory to the Nash Equilibrium.

In this case of sustaining innovation, the evolutionary pathway of production technology will not escape the technological boundaries or production possibilities frontier, without some source of new knowledge gained from a source outside of the firm being applied in the firm.

In other words, codified knowledge leads to asexual innovation because humans are not sharing insights and imaginations about the future.

Richard Nelson and Sidney Winter, in An Evolutionary Theory of Economic Change, (1982), describe the diffusion of codified knowledge as "learning-by-doing."

Applying the codified knowledge in learning-by-doing, creates new technology that "evolves from the old...one round of technology lays the foundation for the next round."

This explanation by Nelson and Winter is a paraphrase of Christensen's sustaining innovation.

The codified knowledge creates better existing products because the production process has become more efficient, without adding new product genetics to the economy.

Some humans, like those that lead large corporations, tend to resist technological change, and try to use social and political mechanisms to control technological change in order to hold on to the status quo of their income distribution.

Both tacit and codified knowledge, as economic assets, are valuable assets to control politically because controlling the creation and diffusion of knowledge is linked to control over the distribution of incomes.

The more that technological innovations can be directed to small genetic changes, which are more likely to be selected by consumers, the less likely the economy will be to break away from equilibrium.

Existing firms, and existing commercial bankers who may have loaned money to technology firms, have a motivation to control the direction of technology in order to maintain their incomes.

The emergence of new tacit knowledge is generally not known to the incumbent firms, which is why, according to Christensen, that the incumbent firms get "blown away" by the introduction of new radical products.

The social forces that resist technology and investments in new products, thus have a evolutionary reason to control codified knowledge by keeping it channeled into very specific genetic varieties.

"Research and development is directed towards shaping and refining knowledge in very specific ways," states Peter Hall. (Innovation, Economics and Evolution: Theoretical Perspectives on Changing Technology in Economic Systems, 1994).

"Research and development is worth doing only if it generates a product or process of commercial value...which both fits with the firm's existing capabilities and meets market requirements (of existing consumers)."

The ability of corporations to control knowledge is a LaMarkian acquired characteristic, but in a negative way.

Controlling knowledge is more like a de-acquired characteristic because when knowledge is controlled, it contributes to the economic extinction of knowledge in subsequent generations.

Control over tacit knowledge means that no new memories are being created for time period two.

This opens the question of how new knowledge in new products are "selected" by consumers who previously had not known about the new products.

Chapter 8.
Consumer Selection and Technology Evolution.

Just like the guesses made by managers of firms about producing a radical new product, and venture capitalists, who are guessing about reaping their profits from an investment in a new product, consumers are also "guessing" about how a new product may "fit" into their welfare function.

The consumer guesses are based upon the notion of the "rational pursuit of self interest," as perceived in the brain of the individual making the decision.

The consumer's brain is filtering thousands of images, trying to come up with something that looks like the new product. The internal images in the brain are comparing the old images with the new external images of the product, in order to come up with a strategy of how to respond to the new thing.

This guessing activity of consumers about novel events is in contrast to consumer behavior in existing markets, with older versions of the product. With old products, the consumer brain has developed rules of thumb making selections, primarily based upon maximizing utility at the lowest cost.

With new products, the selection behavior of consumers is not a priced-based activity, it is an insight-imagination, mental-guess-based activity. Prices for the future market of the product have not been established.

Much of what happens in the theory of technology evolution depends on what the consumer does when she first "sees" the new product. When she first "sees" the new product, she "imagines" how the new product may "fit" with her expectations of the future.

The seeing and imaging part of the explanation is a biological function that occurs in the brains of each consumer, when they are confronted by a new thing.

In the sequence of events leading to technology evolution, the guesses being made by consumers influence the guesses made by both venture capitalists and producers about what consumers are going to do when they first see the new product.

New markets are created when the mental activity of guessing about the future, via the insight-imagination process, creates the future markets as the present market.

Future markets come slowly into the present through a slight opening in time when guesses about the future between consumers, producers, and investors all line up and confirm the new reality.

The evolution of technology is primarily a market adaptation event, and not primarily a technology innovation, or a technology production, event.

According to Wroe Alderson, (Dynamic Marketing Behavior, 1965),

> Suppliers and consumers are engaged in a double search and each is providing the other with clues to guide their search. Consumers specify their needs or, in the case of shopping for goods, partially specify their needs in advance of entering the market. Suppliers identify their products and invite consumers to buy them to see whether the products meet the specified needs.

For existing products, with existing prices, consumers can search for the price that equates their marginal utility of consumption of the last unit consumed with their marginal cost curve, in order to derive maximum rational utility at their exact level of income.

Alderson notes that in the real world,

> consumers are frequently heard to remark that they did not get their money's worth out of a given purchase. Hesitation about buying a product in the first place undoubtedly means that they are trying to estimate this expected value.

The fear and risk of "not getting your money's worth," causes the consumer to hesitate in selecting a radical new product.

In other words, it is not just the single consumption of a new product, at an immediate point in time, which the consumer is evaluating in making the decision to buy.

Consumer risk in buying a new product that may not "fit," is causing Alderson's imaginary consumer to hesitate. The selection of the new product may affect the consumer welfare into the future, and some decisions are irrevocable.

The guessing and search function, in Alderson's scenario, is a separate mental activity from the selection function.

The selection function is the force that moves the existing market towards a new market.

As Alderson notes,

> If strongly motivated problem-solvers (i.e. searchers) face each other in a discrepant market, it can never be cleared but only moves in the direction of that equilibrium state...Traditional economics has no place for marketing effort because saleable products were sold simply by raising or lowering the price to equate supply with effective demand.

In sustaining, asexual innovation, suppliers are searching for customers who will buy their newer, improved, second generation product, at a price that is higher than the first generation product.

The first generation product drops in price and moves down into the lower income consumer market, and as the first generation product moves down in price, it is on its pathway to technological obsolescence and extinction.

In a genetically diverse product environment, with high rates of genetic crossover, an individual consumer is more likely to take a "risk" on selecting the new product with an unproven technological track record.

The consumer's willingness to select a new product is based upon a social environmental condition that concerns the level of social class incomes in the genetically diverse environment.

In sustaining innovation, the more the newer improved product "looks" like the older version of the parent technology, the less time the consumer searches, and the greater the probability that a higher income consumer will select the newer, improved version of the product that is slightly different.

The wealthy individuals possess a higher marginal utility for innovative products, with a greater propensity for taking a risk on the untried product when it first hits the market

They have more income to take a risk, and they have "learned-by-doing" in buying products that looked slightly different than the products that they had seen before.

The risk of selecting an innovative product for the upper income consumer is less than the loss associated with opportunities foregone, if they do not buy the higher priced good.

Poor people generally have a limited experience in selecting genetically new products, and are risk-averse to spending money on a product that may not fit.

The consumer selection behavior on product selection feeds back information from the final demand market to the intermediate demand market of producers about what kinds of new products may work.

It is the market selections made by upper income individuals who have both the higher marginal propensity to consume, and the higher discretionary income to buy the higher priced products, that may cause a new niche market to form.

It is primarily changes in the intermediate demand market that causes a new market niche to form. The major contingency in technology evolution is whether the new product has the potential to create new complementary products in the input output supply chain in intermediate demand markets.

The reason for the primacy of the intermediate demand markets is related to their ability to diffuse knowledge in the production side of the technological community.

Technology evolution is primarily a market event that depends on what upper income consumers do when they first see a new product. What the upper income consumers do affects the creation and diffusion of knowledge in the intermediate demand markets.

Consumer demand in the existing market is acting as the selection mechanism, and in the case of product evolution, this would be the case of adaptive, selection-induced mutation.

When consumers see a new advanced product that looks similar to the old product, but performs somewhat better, technologically, they are more likely to "select" that product than a product that is radically dissimilar.

The likelihood of consumer selection of the new sustaining innovation product goes up in a market environment where upper income consumers are confronted with the appearance of new product phenotypes that are slightly different than the products that they have already seen.

The probability of selection goes down in the case of radical innovation, if the consumer has never seen the technological attributes of the new product.

The more the new product is different genetically from the first generation product, the less likely consumers are to select it because their brains are searching for how the novel situation "fits" with the prior experience.

With radical new products, the consumer mental rules of thumb do not work.

However, the new product selected by consumers has the greatest probability of passing its genes along to the next generation product, in an asexual, adaptive and directional pattern of heredity.

The product selected by consumers has the greatest probability of passing its genes along to the next generation product, in an asexual, adaptive and directional pattern of heredity.

It is positive and directed selection, with genetic technological mutations accumulating in the technological characteristics of products that are being selected by consumers in the existing market, while those products not being selected in the market are eliminated from the mutation process.

Technological closeness in products is analogous to closeness in amino acids in the DNA structure of the folded proteins.

In sustaining innovation, the fate of the original asexual product mutation is altered and affected by the appearance of a slightly superior mutation, which ultimately replaces the original parent product.

The subsequent product will be replaced by some slightly better future product innovation, until the entire repertoire of technological genetic diversity within the economy is used up.

The consumer selection process for radical new products is different than sustaining innovations because radical products do not have any existing preferences.

"Somehow," states Berten Martens, "a method has to be found to account for the emergence of ex nihilo preference arguments for new goods."

Martens envisions a market economy that represents "...a third tier of evolution above Darwinian survival and Lamarckian learned adaptation."

Martens theory of ex nihilo consumer preferences depends on the product "heuristics," of the new product. (The Cognitive Mechanics of Economic Development and Institutional Change, 2004,).

Martens' use of the term "heuristics" has the same meaning as "technological characteristics" as it was used earlier in this book in the context of applying the biological metaphor to asexual, sustaining innovation.

Martens states,

> A new or innovative good will be preferred if its total characteristics vector yields a higher level of consumer satisfaction for the same budget outlay... new goods provide an original recombination of product characteristics.

Martens suggests that the potential shift in consumer preference from existing older products to newer technological versions of that older product can be analyzed with the use of Bayesian statistical probability theory which would indicate the shift in consumer demand as new products emerge which contain properties and characteristics of old products, only better.

The process that Martens is describing is technological evolutionary heredity, based upon genetic technological mutation.

In the case of asexual technology innovation,

> Innovation builds on existing preferences for characteristics and provides only an original or enhanced (re) combination of a bundle of characteristics.

In other words, the reason consumers select a new version of an existing product is that the older consumer preferences transfer over to the new product. The slightly different product has a built-in set of consumer preferences that influence consumer selection of the newer product.

Consumers selecting offspring products rather than taking a risk on radical genetically new products create stability in market demand in asexual mutations. The stability created by asexual product innovation does not threaten the existing distribution of income.

The consumer selection process and habits for sustaining innovation create the conditions for equilibrium, which leads to technological stasis, at the Nash Equilibrium.

In a politically-inbred economy, the elites manipulate political rules to obtain legacy income associated with their prior capital investments, thus killing the food supply of knowledge that fuels endogenous technological evolution.

The product selection of consumers leads to, or "causes," variation in technology to decline as a result of consumers selecting existing products. The selection of asexual products leads to homogeneity in both ideas and innovation.

Applying the biological metaphor, homogeneity in expectations looks something like an economy with little genetic variety in a short period of time.

Homogeneity is social values could be translated into the idea that many mental images, in both the production side of the market and on the consumer side of the market, are the same.

Everyone is thinking alike.

Consumers do not have new products to choose because of political manipulation of the rules that have the effect of eliminating the rate of investment in radical new products.

Mancur Olsen describes the emergence of social resistance to further innovation because the existing status quo distribution of income satisfies the most recent generation of product innovators.(The Rise and Decline of Nations: Economic Growth, Stagflation, and Social Rigidities, 1982).

New knowledge on the part of consumers to understand how new products may fit, and new firms, which may divert income from the existing distribution of income, threaten the status quo local attractor point stability.

The political and social resistance to radical innovation cuts off the creation and diffusion of new knowledge, leading the economy to a Nash Equilibrium with a low level of aggregate demand.

Once the decline in demand for existing technologically obsolete products occurs, it becomes very difficult to re-generate the sources if new knowledge.

The declining rate of new knowledge kills off the channels of information flows within the economy, which are essential for the process of knowledge creation and diffusion.

Charles Kindleberger, in World Economic Primacy, noted how a certain set of cultural values tended to favor an attitude towards technical innovation.

He characterized this attitude as the

> ...capability and will of individuals, companies and governments to break free of existing habits, perceptions, institutions, and task allocations, in order to revise them in light of constantly changing circumstances and developments.

In contrast to the stable equilibrium of asexual innovation, the positive social attitudes towards two-parent sexual genotype crossover, leads to substantial risk of income loss by those who derive the greatest amount of income from the continuation of the status quo.

If consumers select a radical new product, new complementary markets may be created to service and support the new product that consumers have chosen to select.

Future intermediate markets create new flows of income, where none had existed before. The new flows of income disrupt the existing distribution of income.

In economics, the "fitness" of the new product phenotype is analogous to adaptation in biology, and adaptation is the economic equivalent of market competition.

Applying the biological metaphor from Kimura to evolutionary economics, "the smaller the difference between two amino acids, the higher the probability that they are selectively equivalent." (The Natural Law of Molecular Evolution, 1983).

In the case of two-parent genetic crossover, Kimura's insight is that if the distance between the genes of the new product are small, then the new radical product has a greater chance of consumer selection.

The small technological distance, however, creates the smallest contribution to the formation of a new future market.

On the other hand, following Kimura, a large difference in the technology genetics of the two-parents, create an offspring product that is less likely to be selected.

The big gap between the genetics of the parents causes the greatest change in the economic environment, possibly causing leap-frog technological change, if high income consumers take a risk in buying the radical new product.

Of all the two-parent products created, only a small fraction will survive the effects of competitive adaptation from existing products.

Garish and Lenski describe this process in biological evolution by contrasting sexual two-parent crossover and asexual single parent innovation. (The Fate of Competing Beneficial Mutations in an Asexual Population, 1998).

They state:

> In sexual populations, beneficial mutations that occur in different lineages may be recombined into a single lineage. In asexual populations, however, clones that carry such alternative beneficial mutations compete with one another

and, thereby, interfere with the expected progression of a given mutation to fixation.

The leapfrog effect occurs when the "best practice" new product genotype is less closely related to the average product genotype than earlier generations of product mutations associated with asexual heredity.

The product technology "leap-frogs" over the existing technology creating a competitive emergency for existing products.

If existing firms and existing products do not immediately respond to the radical new technology, they will not be able to overcome the growing technological gap.

The leapfrog mutation in product technology is the precursor event that precedes a potential product market bifurcation. The product market bifurcation occurs as a small niche market for the new product is formed, as a result of a growing consumer selection demand.

Saviotti explains, "Niche theory predicts that the number of niches that may be created in a given habitat is proportional to the size of the habitat." (Technological Evolution, Variety and The Economy, 1996).

Saviotti continues,

> The greater the span (in technological diversity) of the habitat, the greater the number of niches that can be created within it. An increase in the range (variety of product technologies) allows a technology to become more specialized and differentiated, thus leading to an increase in both production technology and product variety.

Under economic conditions where a product crossover in technology between two species occurs, if consumers favor that product, the probability for market bifurcation exists.

The bifurcation means that the market characteristics of the older products have been replaced by the new niche market.

The choice of consumers however, is only one necessary condition for market bifurcation in demand from the old patterns of demand.

As Johansson and Andersson correctly note, the bifurcation in market demand depends on the creation of complementary supply chain markets, which support the new product, while at the same time, the complementary markets speed up the rate of technological knowledge creation, via brand new horizontal information flows. (Theories of Endogenous Regional Growth: Lessons For Regional Policies, 2001).

The concept of the Hopf Bifurcation, in evolutionalry biology, suggests where to look for the imaginary consumers who have formed ex nihilo preferences.

The search for the potential new market begins at the intersection between the technologies of the two parents. Just as it would begin in Darwin, at the edge of the biological environment, where two species may coexist.

Part of the market demand of the imaginary consumers is probably located with the technology market of one parent, and part of the future potential market demand is located in the market of the second parent.

The more radical the product, the greater the effort to create *ex nihilo* preferences via marketing, so that consumers can see how the product fits, even at the higher initial prices compared to any other product on the market.

Only a certain type of social/institutional structure supports the environment for niche market creation.

From the biological evolutionary metaphor perspective, heterogeneity in knowledge and consumer expectations supports the appearance of new technological knowledge

In a heterogeneous environment, everyone is not thinking alike as a result of this new technical knowledge. Different brains are imagining different futures, with different technology combinations, at different prices.

If their imaginations about the future are realized, they can imagine how their incomes would change for the better.

In the case of changing incomes, brought about by changing technology and new prices, the environmental conditions would be ripe for an even bigger technological bifurcation, called the emergence of an entirely new future market.

If high income consumers select the new radical product, and a niche market develops, it is possible that a future market bifurcation would create a new distribution of income.

Markets emerge when income distributions change, and incomes will not change without radical technological innovation.

Income competition is the basis of technological innovation, and in that economy, the future market looks nothing like the past market.

Chapter 9.
Income Distribution and Technological Evolution.

During his one-month visit to the Galapagos Islands, Darwin noticed that the beaks of one species of birds on the islands were slightly different.

The birds seemed to have the same parent lineage, but their beaks had been adapted, over time, to better "fit" the specific environment of each island.

In 1859, Darwin published "On the Origin of Species," where he explained how a new species originated as a result of random genetic variation and natural selection, caused by adaptation.

The favorable adaptations of Darwin's bird beaks were selected, by nature, over many generations, until they all branched out to make new species. Their beaks had adapted to the type of food they ate in order to fill different niches on the Galapagos Islands.

Their isolation on the islands over long periods of time made them undergo speciation.

Applying the biological metaphor to the origin of new product phenotypes, the specific interindustry supply chains for products serves as the economic "environment" of both sustaining and radical technology evolution.

The interindustry relationships supply the food of knowledge to the economic birds in the environment so that they can create new species.

Political control over the food supply limits the technological possibilities frontier for new knowledge creation and diffusion.

In Darwin, the competition between species is over food. As Darwin stated,

As natural selection acts by competition for resources, it adapts the inhabitants of each country only in relation to the degree of perfection of their associates"

In the evolution of technology, the competition is over income between incumbent species and new product species. In the survival of the economic fittest, the incumbent firms attempt to kill the new species, before it has a chance to survive.

Only one type of technology evolution leads to the creation of new future flows of income.

Radical new products create new interindustry environments, when the product market niche bifurcates into a new product market. The new bifurcated product market creates income, where none had existed before.

The two-parent cross breeding causes a new strain of genetic technological phenotypes, when the alleles of the two parents are crossed.

Following the insights about inheritance of Gregor Mendel, (Experiments in Plant Hybridization, 1865.), radical new product technological genes are inherited from two parents.

In the first application of the metaphor, each technological trait of the two parents is passed on unchanged to the offspring product.

In the second application of the metaphor, the offspring product inherits one allele from each parent for each technological characteristic.

In the third application, some alleles may not be expressed in the first generation of the new radical product, but those latent, non-expressed genes but can still be passed on to subsequent generations of products who share the same parent lineage.

The third application explains how an initial new product created from crossbreeding, can undergo subsequent sustaining innovations.

The technological gene pool is increased from the two-parent crossover, and those non-expressed genes are the basis of sustaining innovations, until the entire repertoire of genes from the cross breeding is used up.

Income distributions change when consumers select the radically new and disruptive products, and the economic structure changes when all of the complementary service and support industries grow up to supply the intermediate goods to produce the new product.

Jakob Schmookler, in Invention and Economic Growth, (1966), explained the essential function of the intermediate supply chains. His insight was that innovations performed in one technological environment could affect technological innovations in other technological environments.

He stated that the technological innovations were "embodied" in intermediate goods purchased by the sector. The purchase of intermediate goods transmitted "knowledge" to industries purchasing their products.

F. M. Scherer, in Inter-industry Technology Flows and Productivity Growth, (1981), described an "interindustry technology flows" matrix along the lines suggested by Schmookler.

Scherer's research on technology innovation provided empirical support for Schmookler's thesis that "imported" or "embodied" R&D, transmitted through the interindustry supply chain, is an important determinant of productivity growth.

In his review of technological innovation, Andrew Reamer, (The Impacts of Technological Invention on Economic Growth, A Review of the Literature, 2014.), complains about the inadequacy of Schmookler, Scherer, and Joel Mokyr's models of technological innovation as having no explanatory value.

Reamer begins his criticism with the inadequacy of the terms disruptive and sustaining innovation.

Reamer then states,

> Each of these three typologies is problematic for the purposes of exploring the impacts of invention on economic growth. Not only is the first (Schmookler), not consistently defined, no one has combed through history to identify which inventions are radical and which are not and systematically explore the relationship among them. The second (Scherer) depends on a notion that macroinventions emerge outside the dynamics of social science —I don't believe this to be so, but in any case it is not a framework amenable to analysis. Mokyr has not developed a comprehensive list of macroinventions nor has he prepared a rigorous testable framework for understanding the relationship between macroinventions, microinventions, and economic growth. While Christensen's approach is useful for understanding the relationship between invention, product markets, and industry structure, proponents have not come up with a definitive listing of disruptive technologies and mapped the relation between them, sustaining technologies, and economic growth.

Reamer's review of the literature did not include Edward Feser's work on how technological affinities within the intermediate demand matrix transmitted technology from one industry to the other industries, who shared similar technology in production.

Feser's work, (National Industry Cluster Templates: A Framework For Applied Regional Analysis, 2000, and A Test For the Coincident Economic and Spatial Clustering of Business Enterprises, 2002.), provides the analytical framework for investigating the region's knowledge creation and diffusion of new knowledge.

Feser's basic hypothesis is that,

> co-located businesses that are in the same production chain, share similarities in intermediate input consumption, technology or worker skill mix and are related through other intermediary institutions or informal means.

The intermediate demand structure serves a double duty of describing both the production technology of the economy, and the institutional social class structure of market relationships.

His analytical framework of an input-output model, modified by factor analysis, describes how both the feed forward and the feedback forces of new knowledge affect technological evolution.

In contrast to Reamer's reliance on "total factor production," Feser's methodology provides a better understanding of how knowledge, as a factor endowment, or asset, of the economy, can also erode over time.

Feser's interindustry matrix, as it changes over time, is a reflection of how new knowledge is created, diffused and, contingently, commercialized in the form of a new radical product.

As Feser explains, "...it is not just the size of the district alone, but social, cultural, and political factors, including trust, business customs, social ties, and other institutional considerations."

Feser notes,

> the clusters represent distinct technological groupings of sectors or product chains...Although the conduits of interdependence between firms extend well beyond supplier linkages, input-output flows provide the single best uniform means of identifying which firms and industries are most likely to interact through a myriad of interrelated formal and informal channels.

In other words, applying the biological metaphor, Feser's method predicts which product parents, who share technological affinities, are most likely to cross breed to create a new radical product.

It is not obvious, or immediately apparent, which parent products share technological affinities. Some of the combinations between products, such as cell phones and digital cameras, could not be analyzed, without Feser's model.

As Feser explains,

> By grouping those firms that are most likely to interact with each other, both directly and indirectly, the clusters reveal relative specializations in the economy in terms of extended product chains (buyer-supplier, import replacement, and entrepreneurship based strategies) as well as technology deployment and cross firm networking initiatives.

The technology clusters that are uncovered by Feser's method are surrogates for the environmental social networks which allow new knowledge to flow within the economy.

Every transaction coefficient in Feser's A matrix is also describing an underlying knowledge transaction. Each time a new coefficient shows up in the A matrix from crossbreeding, the coefficient is describing the creation of new knowledge.

The higher the density of technological affinities in the matrix, the faster the rate of knowledge contagion, as the disease of new knowledge rapidly infects those knowledge-bearing units closest to the outbreak.

The traditional Mendel genetics heredity matrix can be re-envisioned as a input output matrix, to describe two parent crossbreeding between Parent Product B and Parent Product C.

In Diagram 17, this adaptation of genetics to technology, each parent product is described as having its own unique intermediate demand matrix, noted as bb, and cc.

Diagram 17. Adaptation of Biology Two Parent Genetic Crossover Matrix to Economic Input-Output Intermediate Demand Matrix.

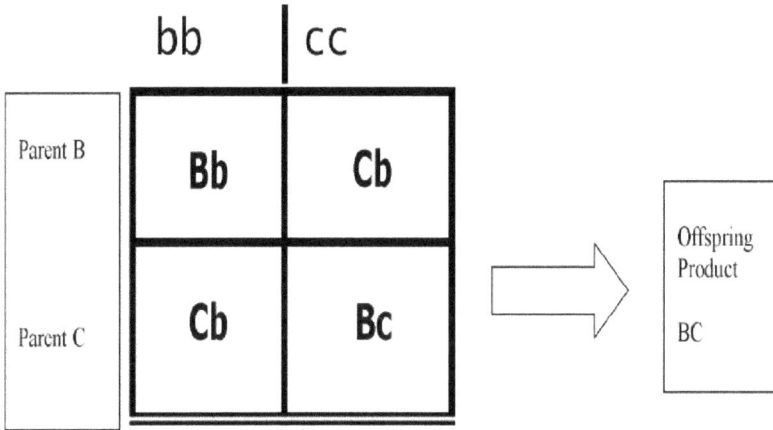

When the two parents crossbreed, components of the new product obtain intermediate inputs from a new matrix bc. The new intermediate demand matrix bc did not exist before the two parent crossbreeding.

Taking a closer look at the new intermediate demand matrix in the form of a traditional Leontief matrix, Diagram 18 describes some of the intermediate supply chains to the new product BC.

In this example, it can be seen that Parent Product C was the dominant genetic parent, and that the intermediate supply chains from C contribute the most factor inputs to new product BC.

The final demand column describes consumer product selection of new product BC.

It is the final demand consumer selections that feed back knowledge from the final demand market to the intermediate demand market about what kinds of new products may work.

In other words, the intermediate demand matrix diffuses knowledge back through all the firms who are supply inputs to Product BC.

The key to understanding technology evolution is to understand how an initial distribution of income from the two parent products B and C changes over time, with the creation of new product BC.

The new distribution of income flows forward, through the new intermediate demand matrix, to the final demand market whenever a firm is paid for its inputs.

Diagram 18. Adaptation of Leontief Input-Output Model to Describe Intermediate Factor Flows to New Product BC.

Inputs (col's) Outputs (rows)	Inputs to New Product BC from Sector 1	Inputs to New Product BC from Sector 2	Final Demand For New Product BC
Intermediate inputs: Sector 1	0.15	0.25	350
Intermediate inputs: Sector 2	0.20	0.05	1700
Primary inputs	0.65	0.70	1100
Total outlays	1.00	1.00	3150

This is an entirely new flow of income, and it is this potential flow of new income that threatens the initial status quo distribution of income obtained by the incumbent Parent Products, B and C, before the two parent crossover occurred..

Under one set of constitutional rules, increasing incomes can be obtained if new markets emerge which have new products that consumers favor over the old products.

New products and new markets may emerge, given a specific configuration of cultural values and laws that favor individual initiative and the appropriation of rewards, based upon individual merit.

In order to restrict the emergence of new products and new markets, the single best strategy of the incumbents to protect their status quo distribution of income is to eliminate the basis of knowledge creation, and to control the rate of capital investment in product innovation.

Both the bankers, and the capitalists who own existing enterprises, have a dependency relationship tied to maintaining the status quo of income distribution that is intermediated by the rate of interest obtained by bankers on commercial loans.

Everyone in the existing status quo arrangement depends on economic and political stability in order to continue to get paid, and nothing in this system benefits from an entrepreneurial change to the status quo arrangement of political power.

Coincidentally, there are powerful financial and economic interests who favor this status quo distribution of income as defining the "right" target level of investment today, just as the communist economists favored the "right" level of investment that yielded their "fair" distribution of income.

In other words, from the perspective of the financial interests of the elite, the political ideological justification for maintaining the status quo is that the existing distribution of income is "fair."

As the competition over legacy income distribution in the dying economy intensifies during the economic meltdown to the Nash equilibrium, the institutional structure of the region undergoes a profound transformation.

In the once, vibrant, open and diverse environment that previously had created and diffused knowledge, income had been growing and distributed according to merit.

The open flows of income in the intermediate demand market, distributed the income, in accordance with the theory of a free competitive economy.

If new genetically superior products do not create new markets, then income distribution will not change in the future.

Without the appearance of new markets, existing goods and services will experience the long run decline in demand, caused by technological obsolescence, as the repertoire of genetics in the economy is used up through sustaining innovation.

For existing products, this decline and obsolescence is analogous to a biological process of brains losing their neuronal connections.

As the economy loses its neuronal connections in the intermediate demand matrix, it becomes increasingly difficult for that economy to re-generate technological innovation.

The state of long-run economic decline is described as Muller's Ratchet of genetic inbreeding. Once the economy goes through a downward ratchet, it is almost impossible to generate future economic growth.

The distribution of income becomes concentrated in the hands of the incumbents, and the common workers that do not benefit from the skewed distribution of income turn to the government for financial support.

Chapter 10.
The Emergence of New Future Markets.

Given a set of initial market conditions, the metaphor of biology asks the question: what happens next when new technology appears in the market?

In sustaining innovation, the existing current market does not contain any means to reproduce itself. The entire repertoire of genetic possibilities for the technological evolution, at the beginning of time period one, is analogous to the market environment at the beginning period of time of the biological evolutionary metaphor.

An increase in technological diversity is the biological equivalent of diversity in genetics, and knowledge is the equivalent of food that gets eaten in a market to provide mental energy to imagine new technology.

New market emergence is a very rare evolutionary event. Sometimes, for long periods of time, as Kimura reminds us, very little happens in evolution at all.

It takes humans a long time to interpret and apply the implications of novelty. It takes even longer for a consensus to form among groups of humans about a common interpretation of the novel event.

New markets emerge as a result of a growing common perception in the brains of humans about the implications of novel market event.

"Consumer wants and preferences are created gradually by means of a learning process involving both producers and consumers," states Koen Frenken, (Technological Innovation and Complexity Theory, 2006).

Consumers learn about new products through the business social networks that feed forward, and feedback, information from the final demand market.

In the biological metaphor, new market emergence implies that former products and former final demand markets have gone extinct.

The emergence of future market environments, after the appearance of a disruptive technology, is contingent upon a certain evolutionary sequence of events taking place.

Applying the biological metaphor, what happens nest in evolution could have been something completely different, given a very slight deleterious mutation.

If another sequence of evolutionary events takes place, income distribution will not change and new markets will not emerge.

The diagram below describes the chronology of events leading to new market emergence.

Diagram 19. Chronological Sequence of Events Leading to A Potential Macro Market Bifurcation .

Appearance of novel radical product from two-parent crossover.	Emergence of small niche market, as consumers select the new product.	Displacement of old product by new radical product. Product market micro bifurcation.	Creation of intermediate demand markets to support production and distribution of product.	Creation of new income flows where none had previously existed.	Profit reinvestment from capital "exit " events into technology trajectory created by new radical product.

If a series of micro market product bifurcations occur, the economic environment may change both its technological production structure, and contingently, its final demand market structure.

The distribution of income that comes after the economy passes through the macro economic bifurcation window is entirely different than the distribution of income that occurred in the earlier period.

After a macro market bifurcation, the economy can never return to the former equilibrium because that market structure has been replaced.

As Joel Mokyr suggests, the "causal chain could thus run from technological success to income and from there, to institutional change, rather than from institutional change to technological change." (The Lever of Riches: Technological Creativity and Economic Progress, 1990).

If the income distribution does not change, then technological evolution will not change. Without new distributions of income, economic growth will stagnate at some lower level of aggregate demand.

The social and political forces that resist a change in income distribution are more powerful than the forces of innovation.

"Technological progress," says Mokyr, "has run into an even more powerful foe: the purposeful self-interested resistance to new technology...Without an understanding of the political economy of technological change, then, the historical development of economic growth will remain a mystery."

The mystery is how any form of economic growth and technological evolution ever occurs, given the allegiance to the status quo.

Given the rare set of circumstances that must coalesce in a chronological sequence of time, it is remarkable that economic evolution occurs at all.

As Mokyr observed, it just so happened in the modern history of economics, that one country would first experience technological innovation, and then economic growth.

Without the addition of new technology, that country would experience economic decline, only to be replaced by some other country that was experiencing technological innovation.

Observing the emergence of a future market is like trying to discover distant planets circling around distant stars. The initial evidence is so small that only the traces of the planet exist in the wobble of the stars that planets revolve around.

In the emergence of a new market, there is a small oscillating moving window of time, a type of threshold Bayesian event, that becomes progressively clearer as the future market becomes the current market.

The Bayesian window is being approached either through upward oscillations, in terms of economic growth, or it is devolving through another type of threshold event of economic decline.

The oscillations towards the future market window become progressively smaller as the transition from the current to the future market is approached because fewer and fewer variables can change, and the rate and scope of change becomes progressively restricted.

Once the current market slips through the Bayesian window, it can never go back in time, and for upward bifurcations, the future market will look nothing like the current market, except in historical terms to note that it evolved out of the current market.

On the downward bifurcation, the economy can stall at a Nash equilibrium for decades at a time.

At the end of the chronological sequence of events on an upward bifurcation, either capital is re-invested into the technological trajectory, leading to future economic growth, or the capital is deployed to maintain the social class structure.

The social class structure affects the birth rate and death rate of ventures because cultural values, like adherence to social rules, trust and loyalty, determines the amount of risk associated with producing a new product.

Most human economic activity is based on the biological ability of humans to anticipate what other humans are going to do, and imagine a course of action to take under different scenarios.

Markets change when humans anticipate the behavior of other humans by imagining the future, and trying out, mentally, what may happen under different hypothetical scenarios.

The behavior choices confronting humans are restricted by cultural values and social customs, which vary by market and by geographical region.

The rational choice set of behavior in future markets being imagined in the brains of humans depends on the freedom to choose, and the freedom to pursue an individual course of destiny, once the choices and decisions about the future have been determined.

In the absence or the presence of a certain configuration of social and cultural values, the risk associated with a future appropriation of profit from a capital gain exit event changes.

The social and cultural values in the ambient environment affect the search, select, and adapt behavior of consumers when they first see the novel product.

In other words, under one set of cultural values, the flows of information, both forward and backwards, are unrestricted from the final demand market for finished goods.

Under another set of cultural values, the incumbents use their capital gains to lock the economy into path dependency in order to restrict the free flow of technological market information.

The initial market niche of the new product determines how fast or slow the information is transmitted, and the selection of new products versus old products, determines how the structure of relationships in the market will change.

At the point of macro market bifurcation, the entire economic technology environment changes, just like the biological environment of the Galápagos Islands, after they emerged from the volcanoes that created them.

The human response to novelty is contingent, and depends on the mental ability of humans to anticipate the future. What humans do in markets when new products are introduced depends on what humans anticipate other humans are going to do when they see the new product for the first time.

New markets are created when the mental activity of guessing about the future via the insight-imagination process creates the future markets as the present market.

Future markets come slowly into the present through a slight opening in time when guesses about the future between consumers and investors are confirmed.

The knowledge creation and adaptive learning in the complex economy creates new markets when current expectations about the future based upon individual guesses are confirmed by collective expectations in future markets.

The new ventures that the venture capitalists create act as a market signaling device, in an information feedback system in the intermediate demand markets that function to coordinate mutually reinforcing expectations between the investors and the consumers.

Stated another way, new markets will not emerge based upon maintenance of the status quo distribution of income.

Evolutionary biology applied to economics can explain how new market demand arises in future markets as a result of the relationships between incomes and technology.

Chapter 11.
The Contribution of Schumpeter to the Theory of Technology Evolution.

The theory of technology explained in this book initially relied on the work of Clayton Christensen in his description of sustaining and radical innovations.

The much more important authority for evolutionary technology in this book is Joseph Schumpeter, (The Theory of Economic Development, 1961).

Schumpeter broke from the herd mentality of neoclassical economics to explain the components of evolutionary economics. And, for his heresy, Schumpeter was shunned by the academic community for decades.

His major insight was that technological innovations displaced existing products, through the gales of creative destruction.

The gales of creative destruction can be placed within the context of the biological metaphor.

Schumpeter was asking the question of how an economy reproduced itself.

His insight was that capital investment created the new firms. His break with the neo classical tradition was that the fund of capital was not derived from price-based exchanges in the sphere of production.

As Schumpeter stated,

> By far, the greater part of it (capital) does not come from thrift in the strict sense, that is from abstaining from the consumption of part of one's regular income, but it consists of funds which are themselves the result of successful innovation, and in which we shall see later recognize entrepreneurial profit.

The profits from the earlier investments are re-invested in the second time period, according to Schumpeter, and the effect of the investment in time period two, is to create new income and wealth in time period three.

The investment process in new firms created new income in the future, which Schumpeter called "purchasing power." The investment in new firms eliminated existing firms.

Schumpeter explained, from the perspective of neoclassical tradition, how the allegiance to the status quo diminished the ability of the economy to reproduce itself.

> The entrepreneur needs capital...to serve as a fund out of which productive goods can be paid for...it is a fund of (future) purchasing power.

The new firms, for Schumpeter, represent the demand for capital, while the capital from the exit events represents the supply of capital.

Schumpeter continued that the economy would stagnate at lower levels of aggregate demand without new capital investment.

> For what business yields interest permanently? The return (from operational profits) of every business ceases after a time, for every business, if it remains unchanged, and soon falls into insignificance...According to our view, the capitalist would first have to lend his capital to one entrepreneur,and after a certain time period to another, since the first cannot be permanently in the position to pay interest.

The source of capital is from previously successful entrepreneurial ventures. The profit generated in the "exit" event of those earlier ventures serves as the source of capital for later ventures.

Schumpeter got the technology evolution sequence of events exactly right:

> It (entrepreneurial profit) attaches to the creation of new things, to the realization of the future value system...Without (entrepreneurial) development, there is no (entrepreneurial) profit, and without profit, no (economic) development...without (entrepreneurial) profit there would be no accumulation of wealth.

According to Schumpeter, the "new thing," created by the entrepreneur is based upon the entrepreneur's application of technology, and comes into the market alongside of the "old thing."

The major dynamic of economic evolution for Schumpeter was the Darwinian competition among species, and the elimination of the weaker species by the stronger species.

He stated,

> ...the new enterprises either completely eliminate old businesses or else force them to restrict their operations.

Schumpeter wrote,

> The (new) demand...causes a rise in the prices of productive services. From this (rise in prices) ensues the withdrawal of good from their previous use...the newly created purchasing power (new incomes) is squeezed out at the cost of previously existing purchasing power...

The newly funded ventures are not subsidiaries or divisions of existing production units.

In Schumpeter, the new enterprise,

> ...does not grow out of the old, but appears alongside of it, and eliminates it (the old enterprise) competitively, so as to

change all the conditions that a special process of adaptation becomes necessary."

These new firms enter the economic fabric at a specific moment in time, and according to Schumpeter, the economic fabric at that moment is seen as a "... circular flow of economic life," that is a closed loop, when capital gains from exit events are re-invested back into the economy.

In Schumpeter's economic model,

> ...sellers of all commodities appear again as buyers in sufficient measure to acquire those goods which will maintain their consumption and their production equipment in the next period at the level so far attained, and vice versa.

Schumpeter described the concept of an economy breaking away from equilibrium as,

> ...a discontinuous change in the channels of the (circular) flow... disturbance of equilibrium, which forever alters and displaces the equilibrium state previously existing economy...These spontaneous and discontinuous changes in the channel of the circular flow and these disturbances of the center of equilibrium appear in the sphere of industrial and commercial life, not in the sphere of the wants of the consumers of the final product.

When Schumpeter explains that the disturbances are in the sphere of industrial life, and not in the market of final demand, he is describing the function of the intermediate demand markets which change forever as a result of technological innovation.

The markets change forever because the economic structure resulting from capital investments are different than the previous economic structure.

Schumpeter states,

> It is not essential that the new combinations should be carried out by the same people who control the production or

commercial process...new combinations mean the competitive elimination of the old...It (the investment process) in new firms explains on the one hand the process by which individuals and families rise and fall economically and socially...In a non-exchange economy, for example, a socialist one...the social consequences (of the investment process) would be wholly absent.

Schumpeter explained that the resistance to the new thing, and new firms, was in the form of political allegiance to the status quo distribution of income.

Existing firms, and existing commercial bankers who may have loaned money to those firms, have a motivation to control the direction of technology in order to maintain their incomes.

The social conflict, as described by Schumpeter, is that bankers depend on stable flows of revenue to repay their loans, and thus, choose only to fund the most cautious and stable of enterprises that come before the selection committee.

The repayment of loans to commercial bankers tends to establish the prevailing rate of interest for the commercial bankers, based upon their assessment of risk. The prevailing rate of interest acts as the target benchmark for capitalists in assessing the profitability of a new investment in a new firm.

As it is in the theory of technology evolution, it is income completion that is the major dynamic for Schumpeter in explaining economic growth.

Schumpeter explained,

> Every individual loan transaction is a real exchange...the exchange of present for future purchasing power...the *control* (italics added for emphasis) of present purchasing power means more future purchasing power (more income) to the borrower.

Under one set of social and constitutional rules, Schumpeter's gales of creative destruction were free to blow.

Under another set of constitutional rules and values, such as monopoly global capitalism, the direction and creation of technology is controlled by a small set of global corporations and global banks, and capital investment is politically controlled and manipulated event that displaces the free competitive market environment.

Late in his life, Schumpeter became pessimistic about the direction of capitalism because he could see how the political allegiance to the status quo was affecting technological evolution, leading to economic stagnation.

Schumpeter's term for monopoly global capitalism was "corporatism." In his evolutionary model, the early phase of entrepreneurial capitalism would lead to corporatism and to a political system disconnected from democracy.

Under monopoly capitalism, the direction and creation of technology is controlled by a small set of corporations, and investment in technology evolution is politically controlled and manipulated event that displaces the free competitive market environment.

Schumpeter's essential point, as it relates to economic evolution, however, was that product technology evolves as a process of disequilibrium, not equilibrium, and that the evolution itself is also an auto-correlative cause of further disequilibrium.

Schumpeter is the first great economist who explained that capitalism can only be understood as an evolutionary process of continuous innovation and creative destruction.

The theory of technology evolution explained in this book places Schumpeter's work into the biological evolutionary metaphor.

The Theory of Technology Bibliography

Alderson, Wroe, Dynamic Marketing Behavior: A Functionalist Theory of Marketing, R. D. Irwin, 1965.

Black, Max, Models and Metaphors: Studies in Language and Philosophy, Cornell University Press. Ithaca, 1962.

Bruno, A.V. and Tybjee, T.T., "The Environment for Entrepreneurship" In Kent, C.A., Sexton, D.L., and Vesper, K.H., (Eds.), Encyclopedia of Entrepreneurship, Prentice-Hall, Englewood Cliffs, N.J., 1982.

Brennan, Geoffrey, and Buchanan, James M., The Reason of Rules: Constitutional Political Economy, Cambridge University Press, London, 1985.

Burger, R., The Mathematical Theory of Selection, Recombination and Mutation, Wiley, 2000.

Christensen, Clayton M., The Innovator's Dilemma: When New Technologies Cause Great Firms To Fail, Harvard Business School Press, Boston, 1997.

Christensen, Clayton, and Raynor, Michael, The Innovator's Solution: Creating and Sustaining Successful Growth, Harvard Business School Press, Boston, 2003.

Christensen, Clayton, M., "The Rules of Innovation," Technology Review, June 2002.

Christensen, Seeing What's Next: Using Theories of Innovation to Predict Industry Change, Harvard Business Review Press, 2004.

Darwin, Charles, On the Origin of Species by Means of Natural Selection, or the Preservation of Favoured Races in the Struggle for Life, John Murray, London, 1869.

Evangelista, Rinaldo Knowledge and Investment: The Sources of Innovation In Industry, Edward Elgar Publishing Ltd, Cheltenham, 1999.

Feser, Edward, J., "Enterprises, External Economies and Economic Development," Journal of Planning Literature, (Vol. 12 #3), February, 1998.

Feser, Edward, J., and Bergman, Ed., "National Industry Cluster Templates: A Framework For Applied Regional Analysis," Regional Studies, (Vol. 34 #1), 2000.

Feser, Edward, J., and Sweeney, Stuart, "A Test For the Coincident Economic and Spatial Clustering of Business Enterprises," Journal of Geographical Systems, (2, 4), 2002.

Forst, Christian, Evolution of Metabolisms: A New Method for the Comparison of Metabolic Pathways, Proceedings of the Annual International Conference on Computational Molecular Biology, RECOMB, New York, 1999.

Garish, Philip J., and Lenski, Richard E., The Fate of Competing Beneficial Mutations in an Asexual Population, Springer, 1998.

Gregoire, Nicolis, and Prigogine, Ilya, Exploring Complexity: An Introduction, St. Martin's Press, 1989.

Guckenheimer, J., Oster, G., and Ipaktchi, A., The Dynamics of Density Dependent Population Models, Springer, 1977.

Hall Peter, Innovation, Economics and Evolution: Theoretical Perspectives on Changing Technology in Economic Systems, Prentice-Hall, 1994.

Hartl, Daniel, Principles of Population Genetics, Sinauer Associates, Sunderland, 1980.

Holland, John H., Emergence: From Chaos To Order, Reading, Helix Books, 1998.

Holland, John, Hidden Order: How Adaptation Builds Complexity, Basic Books, 1996.

Hoyle, Fred, Mathematics of Evolution, Acorn Enterprises, Memphis, 1987.

Hughes, Austin, Adaptive Evolution of Genes and Genomes, Oxford, New York, 1999.

Johansson, Börje, Karlsson Charlie, and Stough, Roger, Theories of Endogenous Regional Growth: Lessons For Regional Policies, Springer, 2001.

Kimura, Motoo, The Natural Law of Molecular Evolution, Cambridge University Press, Cambridge, 1983.

Kindleberger, Charles, P., World Economic Primacy World Economic Primacy: 1500 to 1990. Oxford University Press, New York, 1996.

Leontief, Wassily, Essays In Economics, Theories, Facts and Policies, Volume II, M. E. Sharpe, Inc., White Plains, 1977.

Leontief, Wassily, "Input-Output Data Base For Analysis of Technological Change," Economic Systems Research, (Vol. 1 #3), 1989.

Leontief, Wassily, The Structure of the American Economy, 1919 – 1939: An Empirical Application of Equilibrium Analysis, Oxford University Press, New York, 1960.

Leven, Richard, Evolution in Changing Environments Some Theoretical Explorations, Princeton, 1968.

Lewontin, R. C., The Genetic Basis of Evolutionary Change, Columbia University Press, New York, 1974.

Mansfield, Edwin, (Editor), et. al., The Production and Application of New Industrial Technology, W. W. Norton and Company, New York, 1977

Mansfield, Edwin, (Editor), et. al., Technology Transfer, Productivity, and Economic Policy, W. W. Norton and Company, New York, 1982.

Marsden, J. E., and McCracken, M., The Hopf Bifurcation and Its Applications, Springer-Verlag, New York, 1976.

Marsh, Sarah J., and Stock, Gregory N., Creating Dynamic Capability: The Role of Intertemporal Integration, Knowledge Retention, and Interpretation, Wiley, 2006.

Martens, Bertens, The Cognitive Mechanics of Economic Development and Institutional Change, Routledge, London, 2004.

McAdams, Robert C., Paths of Fire: An Anthropologists Inquiry Into Western Technology, Princeton University Press, Princeton, 1996.

Mendel, Gregor, Experiments in Plant Hybridization, Journal of the Royal Horticultural Society, 26: 1–32. Published 1865, reprinted 2009.

Mitchell, Melanie An Introduction to Genetic Algorithm Mitchell, MIT Press, 1996.

Mokyr, Joel, The Gifts of Athena: Historical Origins of the Knowledge Economy, Princeton University Press, Princeton, 2002.

Mokyr, Joel, The Lever of Riches: Technological Creativity and Economic Progress, Oxford University Press, New York, 1990.

Muller, H.J., The Relation of Recombination to Mutational Advance, Mutation Research/Fundamental and Molecular Mechanisms of Mutagenesis, 1964.

Nelson, Richard, and Winter, Sidney, An Evolutionary Theory of Economic Change, Belknap Press, Cambridge, 1982.

Nelson, Richard B "Technical Change As Cultural Evolution," in Ross, Thomson, (Ed.), Learning and Technical Change, St. Martin's Press, New York, 1993.

Olsen, Mancur,The Rise and Decline of Nations: Economic Growth, Stagflation, and Social Rigidities, Yale University Press,1982.

Reamer Andrew, The Impacts of Technological Invention on Economic Growth, A Review of the Literature, 2014.

Rosenberg, Nathan, The Emergence of Economic Ideas: Essays in the History of Economics, Edward Elgar Publishing, Cambridge, 1994.

Rosenberg, Nathan, Inside the Black Box: Technology and Economics, Cambridge University Press, Cambridge, 1982.

Rosenberg, Nathan, Landau, Ralph and Mowery, David, Technology and the Wealth of Nations, Stanford University Press, Stanford, 1992.

Saviotti, Pier Paolo, Technological Evolution, Variety and The Economy, Edward Elgar, Cheltenham, 1996.

Scherer F. M., in Inter-industry Technology Flows and Productivity Growth, Review of Economics and Statistics 64(4):627-34 · February, 1982.

Schmookler, Jakob, Invention and Economic Growth, Harvard University Press, 1966.

Schumpeter Joseph, The Theory of Economic Development: An Inquiry into Profits, Capital, Credit, Interest, and the Business Cycle, Harvard University Press, 1934.

Smith, John Maynard, Evolutionary Genetics, Oxford University Press, 1998.

Stankiewicz, Rikard, and Carlsson, Bo, Technological Systems and Economic Performance: The Case of Factory Automation, in Technological Systems and Economic Performance: The Case of Factory Automation, Springer, 1995.

Temin, Peter, "Entrepreneurs and Managers," in Higonnet, P., Landes, D. Rosovsky, H., (Eds.), Favorites of Fortune, Harvard University Press, Cambridge, 1991.

Van Graaff, Jan, Theoretical Welfare Economics, Cambridge University Press. Cambridge, 1957.

Woese, Carl, The Genetic Code: the Molecular Basis for Genetic Expression, Harper, 1967.

Zhang, Wei-Bin, Capital and Knowledge: Dynamics of Economic Structures With Non-Constant Returns, Berlin, Springer-Verlag, 1999.

Ziman, John, Technological Innovation As An Evolutionary Process, Cambridge University Press, Cambridge, 2000.

About Laurie Thomas Vass.

Laurie Thomas Vass is a North Carolina regional and constitutional economist.

Vass is a graduate of the University of North Carolina at Chapel Hill and has an undergraduate degree in Political Science and a Masters degree in Regional Planning.

She was a professional money manager with her own investment advisory firm for 30 years, and was cited by Peter Tanous, in The Wealth Equation, as one of the top 100 private managers in the nation.

She is the inventor and holder of a research method patent on selecting technology stocks for investment accounts.

Vass is the author of ten books, and over 100 scholarly economic articles on the Social Science Research Network author platform. She is currently ranked in the top 1.3% of over 420,000 economic authors, worldwide, on the SSRN platform.

Her articles on the SSRN platform are available for free.

She is a student of North Carolina history and public policy, and her books and articles about the state are archived in the Carolina Collection, at Wilson Library, at UNC.

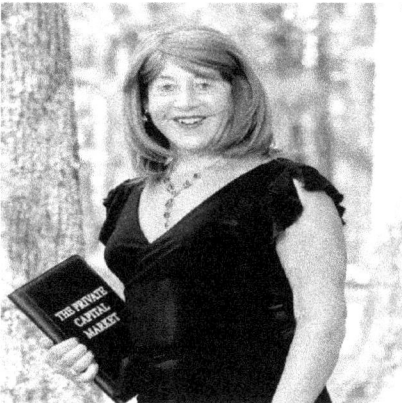

www.ingramcontent.com/pod-product-compliance
Lightning Source LLC
Chambersburg PA
CBHW060319220326
41598CB00027B/4374

*9 7 8 1 5 1 3 6 4 3 9 6 0 *